STRATEGIES FOR PROBLEM SOLVING

TO OUR CHILDREN
Carol, Nicholas, Andrew, Vivienne and Mark.

ABOUT THE AUTHORS
Kaye Stacey (now at Department of Education, Melbourne University) and Susie Groves (Victoria College - Burwood) worked together for ten years in the Department of Mathematics and Computing at Burwood. Their first experiments in teaching mathematical problem solving were for trainee primary and secondary teachers, but these gradually expanded to include work with children in primary and secondary schools and in-service activities for teachers. They both have doctorates in pure mathematics and are both married to mathematicians.

The art work and cover design are by Peter Giles. A brief description of how the pattern on the cover can be obtained is given in Lesson Block 8H.

STRATEGIES FOR PROBLEM SOLVING

LESSON PLANS FOR DEVELOPING MATHEMATICAL THINKING

by

KAYE STACEY

with

SUSIE GROVES

Latitude Publications

The Publishing Division of Latitude Media & Marketing Pty. Ltd.

Published by Latitude Publications
Publishing Division of
Latitude Media & Marketing Pty. Ltd.
41 Martin Place, Glen Waverley,
Victoria, 3150, Australia.

First published by VICTRACC LTD. 1985
Reprinted 1986
Reprinted 1988
Reprinted Latitude Publications 1990

ISBN 0-9589000 2 7

Printed in Australia by
Allanby Press Printers Pty. Ltd.
Camberwell, Victoria.

Copyright © 1985 by K. Stacey and S. Groves-
Limited Reproduction Permission. The authors and publisher hereby grant permission to the teacher who purchases this book, or the teacher for whom this book is purchased, to reproduce up to 100 copies of the student booklet and all Worksheets in this book for use with his or her students. Any further duplication is prohibited.

FOREWORD

We began to develop this series of lessons in 1982 in order to investigate three questions:

> What problem solving strategies can junior secondary students <u>profitably</u> learn?
>
> How can they be taught?
>
> What support materials do teachers need for this task?

The project has been enjoyable, exciting and rewarding because we have worked in co-operation with many teachers and children who have given us enthusiastic support, sound suggestions and kindly criticism. The success of these lessons reflects their constant involvement. In particular we wish to thank Mrs. Barbara Fary, Mrs. Ruth Bunyan, Mrs. Kathy Popper and Mr. Rod Smith, who trialled the "first-draft" materials, and the Year 7 and 8 teachers of Presbyterian Ladies' College (Burwood) who allowed us free access to their classrooms. Mr. Bob Cocking and Mr. Terry Beeby of the Department of Mathematics and Computing (Victoria College - Burwood) assisted with visiting trial schools.

First experiments in writing lesson plans were made easier because we were able to draw upon the experience of the "Skills and Procedures of Mathematical Problem Solving" project (Polytechnic of the South Bank, London, 1980) and use material produced and trialled by it. We would like to thank the director, Dr. Leone Burton, for her generosity in allowing us to adapt the booklet "Helpful Hints For Problem Solvers" to make Lesson Block 7A.

In attempting to teach problem solving strategies, it is not our intention to reduce problem solving to a routine. Rather we hope to broaden the experience of children so that they can tackle problems sensibly, using all their knowledge and creativity. Our work in problem solving, with pre-service teachers and school children, has always been based on a strong philosophy of what mathematics is and, therefore, what children learning mathematics should learn about. We believe mathematics is something that is done, rather than passively learnt. The process of doing mathematics - of thinking mathematically - is enjoyable, it exercises creativity and imagination, and it is within the capacity of almost all children. Good mathematics is not necessarily hard. Children taught by exposition and text book exercises alone see little of the exciting side of mathematics. It is our hope that this book will give teachers a starting point from which they and their pupils can go on to explore mathematics, reflect upon how it is done and thereby become better at it.

It is our special pleasure to thank Mrs. Ann Evans for her constant encouragement and interest and for her excellent typing. We also thank the other secretarial and technical staff of the College, who have always been extremely helpful and professional, and George Warren, the Executive Officer of VICTRACC LTD, who has managed the production.

Kaye Stacey and Susie Groves
Burwood
January 1985

CONTENTS

HOW TO USE THIS BOOK		1
PRACTICAL TEACHING TIPS		5
ABOUT THE STRATEGIES AND HINTS		8
THE LESSON PLANS		13
7A	HELPFUL HINTS FOR PROBLEM SOLVERS	15
7B	LINE DESIGN	27
7C	PATTERN SPOTTING	35
7D	UNDERSTANDING THE QUESTION	42
7E	USING PATTERNS	50
7F	EXPLAINING WHAT YOU DO	59
7G	SPIROLATERALS	64
7H	GENERALISING	72
8A	LEAPFROGS	81
8B	TRIAL AND ERROR	89
8C	FOUR-CUBE HOUSES	94
8D	FIBONACCI NUMBERS	102
8E	CRYPTARITHMS	111
8F	CARD TRICKS	118
8G	STAIRCASE NUMBERS	125
8H	WHAT IF ... ?	130
SUMMARY OF HELPFUL HINTS BOOKLET		136
GUIDE TO THE THEMES AND MATHEMATICAL ACTIVITIES		Inside back cover

HOW TO USE THIS BOOK

UNDERLYING IDEAS

These lessons on problem solving are designed to help children use mathematics they have already learnt to solve non-routine mathematical problems. One of the strongest arguments for teaching and learning mathematics is that it is a useful subject, but it is genuinely useful only if our students can apply it outside the narrow realm of textbook exercises. Opportunities for learning how to do this rarely occur within an expository mode of teaching, entirely aimed at the mastery of specific skills and content areas. To overcome this, teachers are exhorted to "do problem solving" and there is now a large body of literature about problem solving in general, the processes involved and possible problems to use.

This book aims to provide a practical synthesis of theory, strategies and problems in a format for immediate use in the classroom, without reducing the creativity and complexity of mathematical thinking to yet another set of routine exercises. Problem solving is an area of mathematics teaching where many teachers are uncertain about what to teach and how to teach it. Both of these aspects receive attention here - the word "Strategies" in the title refers both to the heuristic strategies useful in tackling problems and to the teaching strategies which can help children acquire them.

It is our belief that three ingredients are essential in any problem solving program:

- **experience in solving non-routine problems**
- **the opportunity to reflect upon these experiences**
- **exposure to some simple problem solving strategies.**

This sequence of lessons has therefore been designed to provide plenty of opportunity for children to tackle intriguing questions, to become engrossed in them and to experience being stuck in a not-too-dreadful way. Selected problem solving strategies are introduced and the lesson plans suggest a variety of teaching strategies to encourage children to think about, discuss, and thereby learn from, these experiences.

In teaching problem solving, as in anything else, the teacher is crucial to the success of the program. It is up to the teacher to create a supportive yet challenging classroom atmosphere, to help students who are stuck (without giving away the answers!) and to show the power of some simple strategies and how they are used. The teacher must draw together the threads, mathematical and strategic, from each problem solving episode, so that children can learn new techniques and refine others. This role is by no means easy, so the lesson plans and the section **Practical Teaching Tips** offer extensive guidance.

TARGET AUDIENCE

The book has been written for teachers of junior secondary mathematics. It is a teacher's handbook - rather than activity cards for children - because children learn more effectively about problem solving if a teacher offers encouragement, thoughtful intervention and discussion about the processes that are being, or could be, employed.

The lessons have been trialled by teachers of year 7 and 8 mathematics classes across a range of abilities. Whilst no knowledge of mathematics beyond primary school is required, the strategies and mathematical processes focussed upon are particularly relevant to junior secondary students. Generalising, for example, is emphasised because of its intrinsic link with algebra. Indeed, many teachers have used these lessons to accompany their introduction of algebra.

Because we feel that it is important for students at every ability level to be given a chance to improve their problem solving skills, activities are designed in such a way that almost all students can expect some success, while more able students can be extended by tackling more difficult **"What if...?"** questions. It is also our experience that most of the problems can easily be adapted so that they are challenging to older students. Indeed, we have used many of these problems with tertiary students.

For example, in Lesson Block 7A, almost all year 7 children are able to find the number of squares on a chessboard, in the sense that they can identify all the different types (sizes) of squares and count how many of each type. Many children will readily see how their method would enable them to calculate the number of squares on square "chessboards" of other dimensions and may be able to explain clearly why their rule works. Counting squares on rectangular "chessboards" and other "What if...?" extensions of the original problem would provide a challenge for senior students, who are concerned with proof and with neat algebraic formulation of their answers.

THE LESSON BLOCKS

The main body of the text consists of sixteen blocks of lessons, each designed to teach an important problem solving skill. The blocks of lessons are independent of each other and take from one to six class periods to complete. Most teachers find it is preferable to take these periods on successive days, but teachers have operated successfully on a once-a-week basis, in which case there is sufficient material here for two years' work.

Teachers can choose to follow the sequence of lesson blocks as it is presented here (thus making a systematic problem solving course) or select particular lesson blocks for occasional lessons. The blocks labelled 7A to 7H have been trialled with year 7 classes, and therefore use slightly easier problems and introduce more elementary aspects of the strategies than lesson blocks 8A to 8H, which have been trialled with year 8 classes. However the difficulty of the problems is largely in the teacher's hands - the suggestions included for adapting the problems ensure that there is plenty of material here to extend students in higher years.

The lesson blocks offer substantial support for the teacher, providing comprehensive lesson notes which include **mathematical details**, a **sample lesson plan**, suggestions for **classroom organisation**, comments on **difficulties commonly experienced by children** and **copyright released worksheets**. Background information on the "process" aspects of mathematics under scrutiny is given in the section **About the Strategies and Hints** while the **Practical Teaching Tips** section gives more ideas on how a classroom atmosphere conducive to problem solving can be developed and maintained.

GUIDE TO THE LESSON BLOCK FORMAT

Theme and Aims

The **Theme** of each block is a strategy, a helpful hint for problem solving or a particular aspect of mathematical thinking. The problems used within the lesson block are chosen primarily as vehicles for illustrating the theme. We have selected, from the many possibilities, themes which emphasise problem solving strategies of central importance and which are accessible to, but not commonly practised by, junior secondary students.

The **Aims** spell out in greater detail what the lesson block hopes to achieve, as well as pointing to other process aspects of mathematics that can be introduced or revised incidentally. These aims are unashamedly ambitious - they cannot be interpreted as behavioural objectives that pupils will attain in a short period of time. Rather they are ways of thinking that can be incorporated into an individual's repertory of cognitive skills only gradually, through experience and reflection upon that experience.

Mathematical Activity

As Polya has observed, problem solving is a practical art, like swimming or playing the piano. Just as it is necessary to go into the water to learn to swim, children learning problem solving must spend much of their time tackling problems. For this reason, the strategies and hints are always introduced and discussed in the context of solving one or more problems. A bald statement of the major problem used in the block is given as the Mathematical Activity. In general we have tried to make these starting problems well-defined, fairly closed and easy in the sense that they can be solved by persistence rather than cleverness. **It is important that all pupils feel they are achieving** - nothing succeeds like success. However, we feel that it is crucial that at least some pupils, some of the time, take the problems well beyond this. Problem solving comes alive for children when they begin to ask their own questions, modifying and adapting the original problem in individual ways. Knowing how problems can be adapted is also the key for teachers to cater for children and classes of varying ages and abilities.

THE LESSON PLANS ARE MUCH EASIER TO READ IF A FEW MINUTES ARE SPENT AT THIS STAGE TRYING THE MATHEMATICAL ACTIVITY AND GOING THROUGH THE WORKSHEETS.

Summary of Lesson Block

As the lesson plans themselves are rather lengthy a summary is included to help teachers select lesson blocks and to provide an overview.

Materials Required and Teacher Preparation

Most lesson blocks include reproducible worksheets at the end of the block. These contain the problems used throughout the lesson block and should be examined by the teacher before reading the lesson plans. Frequently, individual copies of the worksheets are needed.

4 Strategies for Problem Solving

Concrete materials and helpful paper are often useful - and not just for concrete thinkers. None of the materials required are exotic - commonly they include calculators, graph (grid) paper, counters, cubes and scissors.

There is one important omission from the list of items for teacher preparation - **to try the problem beforehand.** A teacher who knows the problem well is better equipped to follow quickly the many individual methods of solution used by class members and is able to adapt the problem should it prove too hard or too easy. This enables teachers to cater for children and classes of varying ages and abilities. Don't discard problems too quickly - instead, learn to adapt them.

Sample Lesson Plans

Of necessity these lesson plans are only examples, written for periods of approximately 45 minutes. Frequently a lesson will take more than one period, but do not let them drag on interminably. Throughout their development they have been trialled by year 7 and 8 classes and they include comments on frequent errors and difficulties experienced by these children. The various lessons within a block are coded in sequence - for example Lesson Block 7A consists of Lessons 7A1, 7A2 and 7A3. While it is not important that the lesson plans are followed closely, it is important that the teacher spends time drawing out the key points contained in each lesson.

Mathematical Notes

This section includes answers for ready reference, background information and discussion of some particularly interesting points. The answers are frequently written using algebra because this is a superb way of communicating with teachers - children would only rarely express their (equivalent) answers in this form. For example, where we may write $y = 3x + 1$, a child in year 7 may say **"it starts at one then goes up by 3 each time"** or **"the answer is always one more than a number in the three times table"**. Several teachers have found situations like this useful motivators for introducing algebra in ordinary mathematics classes.

Worksheets

Ideally, teachers should select from amongst our problems and adapt others to suit their own classes. As we do not live in an ideal world we have supplied the worksheets - please feel free to modify them before reproduction. The worksheets which accompany a lesson are labelled in sequence - for example the worksheets for Lesson 8D2 are labelled Worksheets 8D2-1 and 8D2-2.

PRACTICAL TEACHING TIPS

Teaching problem solving is a new and unfamiliar task for many teachers of mathematics. The normal teaching strategy - exposition, illustrative examples, routine exercises - is inappropriate when the emphasis switches from learning how to do standard tasks to learning about the process of exploring and applying mathematics.

In offering these practical tips it is not our intention to imply that there is only one way to teach problem solving successfully. Teachers need to experiment in order to be able to use elements of their own teaching style to best advantage.

PREPARATION

Please do the problems yourself. Nothing can replace the teacher's first hand experience of doing mathematics.

THE TEACHER'S ROLE - IN SUMMARY

It is up to the teacher to

- **help children accept the challenges:** a problem is not a problem until you want to solve it.
- **build a supportive classroom atmosphere** in which children will be prepared to tackle the unfamiliar and not feel too threatened when they become stuck.
- **allow children to pursue their own paths** towards a solution and assist them when necessary, without giving the answers away.
- provide a framework within which **children can reflect on (i.e. think about, discuss and write about) the processes involved** and thereby learn from experience.
- **talk to the children** about the processes involved in doing and using mathematics, so they can build up a vocabulary for thinking and learning about it. Children learn much more effectively when the teacher draws their attention explicitly to the strategies and processes involved.

COPING WITH INDIVIDUAL DIFFERENCES

This is no harder and perhaps even easier than in other classes. **It is important that every child feels successful:** nothing succeeds like success. Achieve this by starting with a limited fairly closed task that all children can do, if persistent, but encouraging and expecting more able children to tackle more open extensions of the problem. For example, in Lesson Block 7A, all children can become convinced that there are 204 squares on a chessboard by counting them, but more able children should find a rule for predicting how many squares there are on a "chessboard" that is not 8 x 8, perhaps even rectangular.

CLASS SIZE

All the lessons have been trialled as whole class lessons, although the demands placed upon the teacher in trying to follow 30 individual lines of thought can be very heavy. **Knowing the problem well** helps here, as does **allowing children to work in groups of 2 or 3** (pooling ideas means they are stuck less often and the teacher deals with more than one child at a time). Alternatively it is helpful to try **team teaching** (two teachers seem to be able to cope more easily with two classes together than separately) or **taking only half the class** for problem solving whilst the rest are otherwise occupied.

GIVING ANSWERS AND HINTS

Answers: Although it is not important that the children learn "how to solve" any individual problem, it is important that problems are not left up in the air indefinitely. We suggest that **most problems should be "finished off" in a class discussion,** even if this occasionally means telling most of the class the answer. In such a discussion mention could be made of some of the unproductive lines followed by class members. It is important to learn from unsuccessful attempts too.

Hints: One of the most difficult tasks the teacher will encounter is striking the right balance between telling children too much and leaving them stuck for too long. It is counterproductive if children just feel stuck, unable to make a move, but it is also counterproductive if the problem is solved for them by someone else.

Hints that are really only encouragement are often all that are needed: "Stuck? Don't worry, what could you try next?" Hints can point indirectly or, less beneficially, directly to problem solving strategies: "What strategies do you think you could try here?" or "Have you looked for a pattern in these numbers?" Direct, context-related hints such as "What do you get if you find the differences between these numbers?" should be used sparingly and "Have you spotted the triangular numbers?" should be saved for dire emergencies.

ASSESSING PROBLEM SOLVING

Lesson Block 7B includes problems that could be used to give a crude measure of what improvement the class is making in problem solving and the notes accompanying can be used as a guide to marking problem solving assignments. If formal assessment is required for the children or for the teacher's record, it is probably better if it comes in a number of components such as progress made towards a solution, quality of explanation, implementation of a particular strategy (if appropriate), tackling an extension of the original problem etc.

DON'T RUSH THROUGH THE LESSONS

Whilst it is desirable to maintain a reasonable pace and not let lesson blocks drag on for too long, it is also important to give students time to explore some problems in depth and to report to other children in the class. There is a lot to learn from each experience.

FINISHING OFF A PERIOD

It is a good idea to finish most periods with a summary of the important points that have been made during the lesson. We have found that teachers usually do not allow sufficient time for this, especially when children are working well.

DISCOURAGE RUBBING OUT

In rough working, discourage excessive preoccupation with neatness and rubbing (or whiting) out. Encourage light crossing out and the writing of quick notes such as "wrong, I forgot to halve the numbers".

PLAN THE LESSON - but not so rigidly that the exciting moment is lost.

WRITTEN WORK

Most teachers are disappointed by the poor standard of pupils' written work and discouraged by the effort required to produce it. Despite this, children should be required to produce adequate reports of half to one third of the problems they attempt. Fortunately, teachers have found that this is one area where children often show early and substantial improvement as they begin to realise how much thought is involved and begin to make decisions about what is, or is not, relevant to a neat solution. **Explaining what you have done so that someone else can understand** is an important skill to develop and it encourages both checking and reflecting. Students learn more if they study problems in depth rather than flitting from one problem to another.

Try to make the task of writing up as easy as possible, encouraging pictures and diagrams where appropriate. We suggest the following format:

Statement of Problem	Copied from board or pasted in perhaps.
All Working	This section belongs to its author: it is not for anyone else to read. However teachers should encourage children to write down all their ideas.
What I Discovered	A clear explanation of what has been found and why it is true, summarising rather than repeating previous working.
What Ideas I Found Helpful	A brief statement of one or more hints or strategies that helped with the problem.

For Lesson Block 7A, in the "What I Discovered" section, a good student might list the number of squares of each size, show that the total is 204, and give a diagram to illustrate why there are 36 squares of size 3 x 3. For the "What Ideas I Found Helpful" a good answer might be "I found grid paper and counting systematically helped."

Students are often confused about what needs to be done neatly and what is rough work, trying to please the teacher rather than think things out for themselves. Sometimes it is useful to adopt a ploy like "Today we are looking for the best diagram."

ABOUT THE STRATEGIES AND HINTS

The theme of each lesson block is a helpful hint, a problem solving strategy or a particular aspect of mathematical thinking. Of the myriad of hints, skills and strategies, we have chosen to emphasise those which we believe are

- widely applicable to mathematical and non-mathematical problems
- not commonly used by junior secondary pupils
- able to be learnt without too much effort by junior secondary pupils.

Some of these strategies and hints are quite simple and children who begin to use them will reap immediate benefit. Others are more subtle and their full power is attained only gradually, after many rich experiences.

STUCK?

Inevitably, if a problem is a genuine problem for the student, he will become stuck and will turn to the teacher for help. Ideally, teachers should aim to offer help in such a way that students learn to develop the resources necessary to help themselves rather than always needing the teacher.

There are three steps in dealing with being stuck:
- recognise that you are stuck and accept it,
- stop panicking,
- do something about it!

The first two steps, although in a sense trivial, are essential preliminaries. Only a student who has conquered feelings of panic can think productively about new avenues to try.

What can they do about it? To prompt further action, we suggest that children learn to ask themselves five questions:

- What do I KNOW about the problem?
- What do I WANT to find out?
- What can I USE to help me?
- Can I make a GUESS?
- Can I CHECK what I've found?

Of course, questions like these are not helpful in themselves until they become meaningful by gathering many rich associations of how they have been useful in the past. Only then can asking the question become a trigger which reminds the student of things to do. For this reason, we hope teachers will frequently use these questions when they help children, pointing out how they can be related to the strategies and hints that are being studied. More details are given in "Thinking Mathematically" (Mason, Burton and Stacey, Addison Wesley, London, 1982) which is a useful resource and background book for teachers.

THE STRATEGIES AND HINTS

Some background information on the hints and strategies that occur in the course, and why they are there, is given below. We begin with the points introduced in "Helpful Hints For Problem Solvers", the booklet accompanying Lesson Block 7A.

Read it - really understand it.

Reading the problem carefully cannot be stressed too much. It applies not just to mathematics but to all questions and instructions. Careless reading lies behind many failures.

Reading must not be passive, but should usually involve writing, drawing and putting the question into your own words. **Careful initial reading** reduces the chance of a wrong mental set developing and **active re-reading** is aimed at locating misunderstandings and missed information. Point out to the children that it is normal (and a good idea) to read the question **whenever** you are stuck, not just when you begin the problem.

Both the questions "What do I KNOW about the problem?" and "What do I WANT to do?" are relevant to careful reading, and they point to the two different types of information to be extracted from a problem statement.

Write down what you do.

We believe that it is a good idea to encourage children to keep writing as they solve problems. Children are often reluctant to write down a good idea and they then forget it or do the same thing twice. Writing down that something has been tried and doesn't work, or is impossible, is just as important as writing down something which is correct - especially if work is carried over from one day to the next. Writing down a plan often forces it to be clarified.

In the introductory booklet we offer the advice MAKE A START - WRITE OR DO SOMETHING. Making a start on a problem is perhaps the most difficult part of problem solving. Do anything to stop the children staring at a blank sheet of paper. They must not be afraid of writing down something wrong or untidy.

Thinking about what you KNOW and WANT is a suggestion offered to get ideas flowing. It is useful to reconsider what you KNOW and WANT (perhaps by re-reading) and to write it down whenever you are stuck.

Work systematically.

Working systematically is a simple but far-reaching habit that can make a dramatic difference to problem solving. At its simplest it means taking things one at a time, setting them out well, numbering pages and generally knowing where you are going. At a more complex level, considering cases in a systematic way assists in finding patterns and it transforms trial and error from a haphazard "you-might-be-lucky" strategy into an effective tool.

Since tables and charts encourage a systematic approach, we hope that they are frequent answers to the question "What can I USE to help me?"

Strategies for Problem Solving

Use something to help you.

There are many simple concrete aids that make solving a problem easier - not just for concrete thinkers. Many can be readily improvised, but some are best provided by the teacher. Grid paper is particularly useful for diagrams and drawing up tables. Where dot paper and isometric paper are useful, masters have been supplied.

As well as concrete tools, children should be encouraged to consider what **mathematical tools** may be useful. The range available varies greatly with previous mathematical background, but includes ways of presenting data graphically and (increasing through secondary school) the use of pronumerals for expressing ideas.

Beware however: children usually cannot apply, to a non-routine question, any piece of mathematics with which they are not totally familiar. There is a definite place for teachers to demonstrate how a recently acquired mathematical technique can be used in a solution, but this is often best done after children have presented their own solutions. Do not push children to use tools they do not understand when the aim is to focus on problem solving strategies.

Look for and exploit pattern.

Finding and using patterns is surely one of the strategies most characteristic of mathematics. We have generally found that children come from primary school with a strong feeling for pattern and a great faith in it. In these lessons we hope to provide them with more techniques for unearthing number patterns and some understanding that patterns need to be checked and verified before they are used indiscriminately. Particular lesson blocks are devoted to spotting patterns, using patterns and articulating patterns, but it is a theme that underlies most of the book.

Use trial and error.

We have chosen to include trial and error as an important part of the course for several reasons. It is a strategy naturally used by young children, but in a haphazard way that is frequently unsuccessful. Some effort in refining it (so that errors are examined rather than despaired of) pays quick dividends. A second reason is that it is applicable to a huge variety of problems and, with the increasing capability of computers, will become an increasingly widespread method. Trial and error embodies an extremely powerful attitude to solving problems that is not encouraged by routine class work, namely: **if you don't know what to do, try something (anything!) and see what happens.** Of course, working systematically is crucial to the success of the technique.

Develop a good system of recording.

In problem solving, students will often deal with real situations that must be codified and put into symbols before questions can be answered. For example, several lesson blocks use strategy games. To analyse such a game it is almost always necessary to write down information, such as the sequence of moves made and the state of play. Finding a good way of recording this information is by no means easy, yet it is fundamental to success in analysing the situation. The notation needs to be simple to write and remember, and it must capture all the essential information as concisely as possible. Making such a transition from a real situation to symbolic form is an essential preliminary step in problem solving, so it figures prominently in several lesson blocks.

Explain what you have done.

When children have finished a problem, or have done as much as they can, they should be encouraged to review their work and produce a neat, coherent account of it. The idea is to write down the solution with reasons, not just "the answer". The request **"so that someone else can understand"** is one way of trying to explain this.

You may find that children regard this as an irksome task because it stops them getting on to new problems, but it is worthwhile persisting. The process of explaining forces a child to reconsider the solution and gives an opportunity to reflect on what has been experienced. Children learn more this way than by only dealing superficially with problems.

As well as explaining "what", students should be encouraged to explain "why". These explanations are informal proofs and, throughout secondary school, teachers should look for a gradual progression in the tightness of these arguments. Proofs do not have to be full of symbols - words are frequently better.

Check your work.

Checking is an obviously important activity that should be done at intervals while working on a problem. However, children are often surprised to learn that checking does not just mean going over arithmetical calculations. It is important not to repeat what has been done but to derive it in a different way. Several distinct aspects of checking are treated in the lessons.

Generalise your work.

The phrase used in "Helpful Hints For Problem Solvers" to suggest to children that they can generalise results is **Ask Yourself "What if ...?."** We think generalising should be emphasised because:

- it is a basic process of mathematics
- it develops a more complete understanding of the results
- it encourages a thorough review of what has been done
- the question you ask yourself is the one you are most motivated to solve.

Generalisations and other "What if...?" questions should not be forced. The most natural ones arise from an understanding of how the first solution really worked or an individual child's curiousity and creativity.

THE LESSON PLANS

THE LESSON PLANS

Lesson Block 7A

HELPFUL HINTS FOR PROBLEM SOLVERS

Theme   **Helpful hints for solving problems.**

Aims   To illustrate what might be involved in tackling a non-routine problem.

To introduce and illustrate some very simple strategies and good habits to use when solving problems. (These are expanded upon in later lesson blocks.)

To provide a model that children can follow when solving problems. (Each child should keep the booklet for use in later lessons. The hints are summarised on the last page for easy reference.)

Mathematical Activity

SQUARES ON A CHESSBOARD

Someone once told me there are 204 squares on a chessboard. Was she right?

Summary of Lesson Block

Lesson 7A1   Doing the 8x8 chessboard problem and reading pages 1 - 5 of booklet.

Lesson 7A2   Explaining the solution for the 8 x 8 case and reading pages 6 and 7 of the booklet. Trying chessboard problem(s) using different size boards.

Lesson 7A3   Trying a **"What if...?"**

Materials Required and Teacher Preparation

Children require individual copies of the 8-page booklet "Helpful Hints For Problem Solvers", which is made by duplicating and stapling the masters given at the end of this lesson block. All children will require plenty of squared (grid) paper. A "graph blackboard" with an 8 x 8 square outlined in colour is useful - if you can draw on it. A real chessboard is useful if some students are not familiar with it. Some younger children may find a small (9 or 16 pin) geoboard and elastic bands useful to aid visualisation. Paper or cardboard cut-out squares in different sizes (say 3 x 3, 4 x 4 and 5 x 5 to match the graph blackboard) are useful.

Before any problem solving lesson, it is absolutely essential that the teacher should have tried the problem. Please do not leave out this step as nothing can replace first hand experience. Read the booklet and relate it to your own solution. Jot down several "What if....?" questions and try to answer one or more of these.

16 Strategies for Problem Solving

SAMPLE LESSON PLAN

Lesson 7A1

1. Explain to children that in this part of their maths they will solve many different types of problems and learn some simple things to help them to become better problem solvers.

2. Turn to page 1 of the booklet and read it through with the children. **Ask for ideas.** (Usually some children will say there are **64 squares** and then someone will notice that the **whole chessboard itself** is also a square. This will lead to noticing **squares of other sizes.**) Read about this on page 2.

3. Ask children to work individually or in pairs to **find as many squares as they can.** Show them the paper available for drawing chessboards. Discuss whether the black and white colouring is important for this problem. Steer children towards counting the bigger squares first - there are less of them and the resulting number pattern 1, 4, 9, ... is more obvious.
   The assumption that squares cannot overlap is common. (Later on this may lead to an interesting **"What if ...?"** question.)
   It is unlikely at this stage that children will be stuck. However, when children are stuck, tell them **not to worry** and explain that people are often stuck when working on a problem. Ask them to explain what they have been doing and try to help them along the lines of point 7 of the booklet summary - **do not tell them how to do the next step,** try to help them use the summary instead.

4. In the last 10 minutes, read through the booklet to the bottom of page 5. Try to relate it to the children's own solution - encourage comments and discussion. Some teachers like to let children colour in the hints in the text as they are discussed.

5. For homework, children can finish finding all the squares on a chessboard.

Lesson 7A2

1. Remind children of the chessboard problem and ask how many squares they found. Explain that you can learn a lot from carefully looking over what you have done.

2. Ask children how many 4 x 4 squares they found and select one or more children to illustrate how they counted them. (Use cardboard 4 x 4 square if appropriate.) Through class discussion, develop the first table on page 7 of the booklet. Point out some of the number patterns apparent in the table. Relate these to the different methods of counting the squares.

3. Read and explain the remaining points on pages 6 and 7 of the booklet, keeping the "What if...?" investigation for the next lesson.

4. To revise "Helpful Hints" and consolidate the 8 x 8 solution, try to find the number of squares on a 5 x 5 chessboard. Do this as a teacher-led discussion but make sure that the children solve the problem - not you. Teachers can divide the work amongst members of the class by asking different groups of children to find the number of 2 x 2, 3 x 3 and 4 x 4 squares. Draw attention to the appropriate helpful hints as they arise while working on the problem.
   (At this stage, do not worry too much if not all children participate in the discussion. However, make sure that all children do something, such as finding the number of 2 x 2 squares.)

5. If there is time ask all children, working individually or in pairs, to find the number of squares on a 6 x 6 chessboard.

6. For homework, ask children to write up the last chessboard problem tried.

Lesson 7A3

1. Read the last part of page 7 of the booklet. Ask for suggestions for possible **"What if...?"**s and write them on chalkboard. Encourage questioning from the class to help formulate suggestions more precisely. After the list is on the board, briefly comment on which ones might be suitable to try - give reasons, where possible.

    Some **"What if...?"**s that children have suggested in the past are:

    What if we double the size of the chessboard - would we double the number of squares in it?

    What if we counted rectangles (instead of squares) on a chessboard?

    What if we counted squares on a rectangular "chessboard"?

    What if we only count non-overlapping squares? (This one takes a lot more formulation.)

    What if we counted triangles in a triangle?

    How many triangles?

2. Ask children, individually or in pairs, to select a suitable **"What if...?"** and attempt a solution.

3. If there is time, ask a few children, who have attempted different **"What if...?"**s to briefly explain them at board. Encourage questions and discussion from the class. You will probably find that many children have difficulty in clearly formulating their problem - this will usually become obvious while they attempt to describe their solutions. Encourage discussion to clarify the problem. Emphasize that this is difficult for everyone.

4. For homework, ask children to write up a **"What if...?"**

MATHEMATICAL NOTES

SOLUTION TO THE ORIGINAL PROBLEM

The solution to the chessboard problem given in the booklet is not the only one. Here is another way, as reported by Ilana Werba, Year 7, Mt Scopus College.

"She could be right. How many squares are inside others? There are 64 individual little squares. There is no more than 1 big square 8 x 8.

I put little circles around the middle of each 2 x 2 square and counted them all at the end. There are 49 so that means all the 2 x 2 squares equal 49.

I put little crosses in the middle square of each 3 x 3 square and counted them all at the end. There are 36 so that means all the 3 x 3 squares equal 36.

18 Strategies for Problem Solving

I worked out that there were three 6 x 6 squares in one row and there were three rows, so there must be nine 6 x 6 squares.

I then worked out that there were two 7 x 7 squares in a row and there were 2 rows so there must be four 7 x 7 squares.

I worked out that each of these numbers followed a pattern, they were all the square numbers. So 4 x 4 must equal 25 and 5 x 5 must equal 16.

Size of square	1x1	2x2	3x3	4x4	5x5	6x6	7x7	8x8
Number of squares	64	49	36	25	16	9	4	1

So I added all these numbers up. The girl was right, there are 204 squares on a chessboard."

SOLUTIONS TO SOME "WHAT IF...?" PROBLEMS.

a) The number of squares on an n x n chessboard is $1^2 + 2^2 + ... + n^2$. This result follows easily from the booklet.

b) The number of rectangles (including squares) on an n x n chessboard works out to be $1^3 + 2^3 + 3^3 + ... + n^3$.

The method of counting the rectangles is similar to the method of counting squares but the adding up is more complicated and produces the intriguing identity

$$1^3 + 2^3 + ... + n^3 = (1 + 2 + ... + n)^2.$$

c) The number of cubes in a three dimensional "chessboard" of size n x n x n is also $1^3 + 2^3 + ... + n^3$. These can be counted directly using a ready generalisation of the method in the booklet.

d) The number of squares (or rectangles) on a rectangular grid is worse!

Clearly, children will not be able to find general results, so restrict their work to treating particular numerical cases. Remember that the basic patterns in the results are still the same. For example, there are twelve 5x11 rectangles in a 7x14 rectangle. This is because there are 3 positions for a side of length 5 along a 7 unit dimension and 4 positions for a side of length 11 along a 14 unit dimension. Thus there are 3x4 positions for the 5 x 11 rectangle.

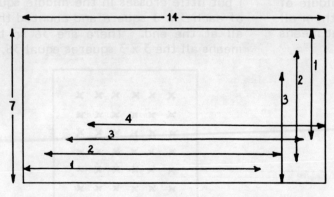

The three positions for a side of length 5

The four positions for a side of length 11

Strategies for Problem Solving   Helpful Hints Booklet

HELPFUL HINTS FOR PROBLEM SOLVERS

Someone once told me that there are 204 squares on a chessboard.

Was she right?

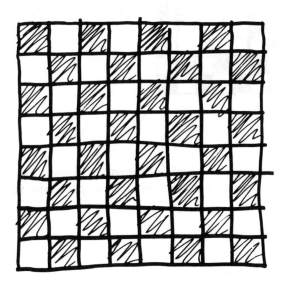

This is the kind of question that makes you think.

Thinking things out can be fun.

Now think for a few moments about how many squares there really are on a chessboard.

THEN TURN OVER

Did you count 8 rows of 8 squares? That is 8 x 8 or 64 squares.

Did you spot the big square round the edge of the board?
That makes 65 squares.

And there are many more squares like this

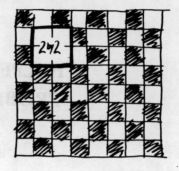

and like this 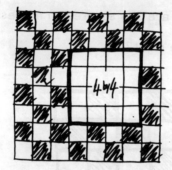 and like this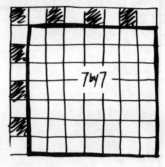

There are a LOT of squares on a chessboard. Finding them all will take you some time - it is a BIG problem. Have another go at the problem for a few minutes. How many squares can you find?

 Working on a problem is like painting a picture or writing a story.

You have to put a lot of thought into a picture or a story to make something good to look at or good to read.

It is the same with a problem. You have to think a lot and you will enjoy doing it.

How do you get ready to paint a picture?

First of all you THINK about what you are going to paint. If you are wanting to paint a house, you look at it as hard as you can, NOTICING EVERYTHING that you want to paint.

It's the same with a problem. To get ready you must

READ IT - REALLY UNDERSTAND IT.

I'll show you what I mean. In the chessboard problem it says there are 204 squares on a chessboard. Did you think it said there are 204 **small** squares on a chessboard?

It's very easy to think a problem says something it doesn't say, so read it carefully.

When you've read the problem carefully

MAKE A START - WRITE OR DRAW SOMETHING.

Even if you can't see what to do straight away, consider what you KNOW about the problem and what you WANT to do.

Then make sure you write it down.

DO THIS NOW IF YOU HAVEN'T ALREADY.

Perhaps you've written something like this:

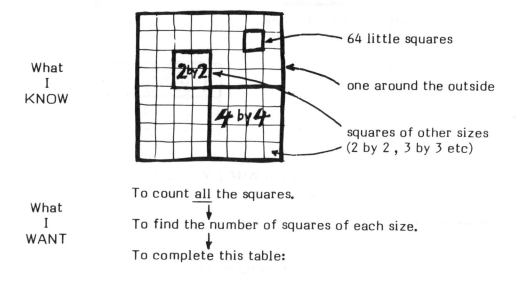

What I KNOW
- 64 little squares
- one around the outside
- squares of other sizes (2 by 2, 3 by 3 etc)

What I WANT
- To count all the squares.
- To find the number of squares of each size.
- To complete this table:

Size of square	1 by 1	2 by 2	3 by 3	4 by 4	5 by 5	6 by 6	7 by 7	8 by 8
Number of squares	64	?	?	?	?	?	?	1

This makes it clear what has to be done. The first step is to find how many 2 x 2 squares there are.

TRY THIS NOW

There are many things you can USE to help you solve a problem. Did you find it took a long time to draw all the diagrams and that your squares got a bit mixed up?

Grid paper and coloured pens can help! For other problems you might use a model made out of cubes or straws. Scissors and paste often help. Sometimes you might use different paper or counters or a graph or a diagram or a table. You might like to use algebra. There are many possibilities.

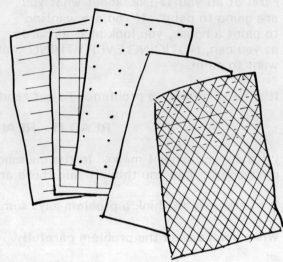

Remember

USE SOMETHING TO HELP YOU.

When I started to count the 2 by 2 squares, I suddenly had the idea that squares could overlap!
So I drew myself this message:

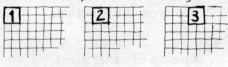
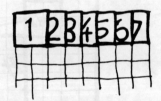

Make it a habit to

WRITE DOWN WHAT YOU DO.

Then you won't forget
- your good ideas
- what you've already discovered
- what you're trying to do.

Another hint: Sometimes you may need to spread out all your working to see if you can find any more ideas. If you are working on separate pieces of paper it's a good idea not to use the backs. DON'T TURN OVER, but take another piece of paper and

NUMBER EACH PAGE.

If you are working in a book and can't take the pages out, try not to draw important diagrams or tables on pages where there isn't much room left. Take a new page instead.

NUMBER EACH PAGE YOU'VE USED NOW

Have you found that there are 49 squares of size 2 by 2?

If you counted the squares higgledy - piggledy then you probably didn't get the right number.

To be sure you've got them all, you need a **systematic** way of counting, perhaps like this:

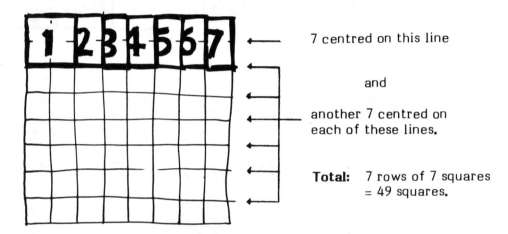

7 centred on this line

and

another 7 centred on each of these lines.

Total: 7 rows of 7 squares = 49 squares.

Problem solving is usually easier if you

WORK SYSTEMATICALLY.

Now try to count the 3 by 3 squares. Work systematically and

DON'T WORRY IF YOU GET STUCK.

After all, it wouldn't be a real problem if you didn't. The main thing is not to panic because there are LOTS OF THINGS you can do about being stuck!

You will learn more about these things later.

TRY TO FINISH THE CHESSBOARD PROBLEM BEFORE YOU GO ON

When you have got an answer (or have gone as far as you can), DON'T PUT THE PROBLEM AWAY. Instead try to

EXPLAIN WHAT YOU HAVE DONE.

Sharing what you have done gives someone else a chance to find out about it. But more importantly, you will find that doing this makes your solution clearer for you too. It also helps you find any mistakes.

NOW TRY TO EXPLAIN YOUR SOLUTION SO SOMEONE ELSE COULD UNDERSTAND

Here is my solution. I hope you can understand it.

How many squares are there on a chessboard?

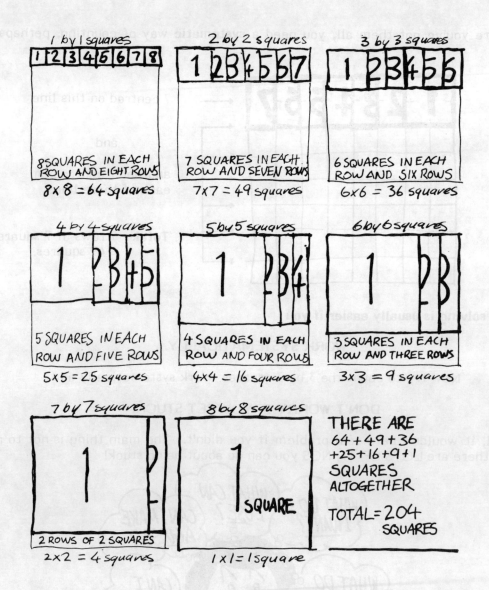

Of course when I first worked it out, my solution was not neat and logical like this one. It took me a lot of work to organize it, but it was worthwhile because I learnt a lot about the problem.

Here are some of the things I learnt.

Firstly I noticed that the number of squares of each size is a SQUARE NUMBER (a number multiplied by itself).

Then I noticed that the **size of the square** and the **number of them in each row** always add up to nine.

Size of square	1 by 1	2 by 2	3 by 3	4 by 4	5 by 5	6 by 6	7 by 7	8 by 8
Number of squares	8 x 8	7 x 7	6 x 6	5 x 5	4 x 4	3 x 3	2 x 2	1 x 1

ALL SQUARE NUMBERS

1+8=9, 2+7=9, 3+6=9, 4+5=9, 5+4=9, 6+3=9, 7+2=9, 8+1=9

After I thought about WHY that happens, I realised that I could use the pattern to work out how many squares there are on a chessboard of any size.

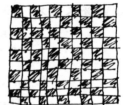

Say I had a 10 by 10 "chessboard".
In a row there are 10 squares of size 1 by 1.

Because 10 + 1 = 11, my pattern predicts that the size of a square plus the number of them in any row will be 11.

Size of square	1 by 1	2 by 2	3 by 3	4 by 4	5 by 5	6 by 6	7 by 7	8 by 8	9 by 9	10 by 10
Number of squares	10x10	9x9	8x8	7x7	6x6	5x5	4x4	3x3	2x2	1x1

So the total number of squares on a 10 by 10 chessboard is

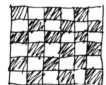

100 + 81 + 64 + 49 + 36 + 25 + 16 + 9 + 4 + 1 = 385.

Now use what you have learnt to find the number of squares on a 6 by 6 "chessboard".

Often, when you've solved a problem you will feel pleased and will want to use what you have found out again. That's why I thought about the 10 by 10 chessboard.

One way of finding another problem to try is to look at the problem again and

ASK YOURSELF "WHAT IF.....?"

In the chessboard problem, I asked "WHAT IF the chessboard was not 8 by 8?"

Think of other "WHAT IF...?" ideas and INVESTIGATE ONE OF THEM.

HAPPY PROBLEM SOLVING!

HELPFUL HINTS FOR PROBLEM SOLVERS

TO START
1. READ THE PROBLEM - REALLY UNDERSTAND IT.
2. MAKE A START - WRITE OR DRAW SOMETHING.
   If you can't see what to do straight away, find
   out what you WANT to do and what you KNOW about it.

AS YOU WORK
3. USE SOMETHING TO HELP YOU.
   Draw a diagram; choose helpful paper; use coloured pens, cubes, counters, scissors and paste, graphs, algebra.....
4. WRITE DOWN WHAT YOU DO.
   Then you won't forget what you did and how you did it.
   Jot down your ideas.
5. NUMBER EACH PAGE.
6. WORK SYSTEMATICALLY.

STUCK?
7. DON'T WORRY IF YOU GET STUCK !
   There are lots of things to do about it. Ask yourself:
   - What do I KNOW about the problem?
   - What do I WANT to do?
   - Can I USE something to help me?
   - Can I make a GUESS?
   - Can I CHECK what I've done?

FINISHED?
8. When you have an answer EXPLAIN WHAT YOU HAVE DONE
   so someone else can understand.
   CHECK your work.
9. ASK YOURSELF "WHAT IF....?" to get ideas for other problems.

Lesson Block 7B

LINE DESIGN

Theme **There are many different ways to do a problem.**

Aims To illustrate that **there are many different ways to do a problem.**

To show children that **making a guess and trying to check it** is a good way to progress if you're stuck. It is important to check using what you know, not what you don't know.

The problem can be administered as a test to provide **a crude measure of the current problem solving ability of the class.** A similar test problem can be done at the end of the year for comparison.

Mathematical Activity

LINE DESIGN

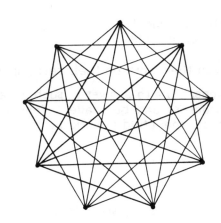

This pattern was made by placing nine dots in a circle and joining each dot to every other dot, except its nearest neighbours. The pattern needed 27 lines.

How many lines would be needed to make a similar pattern using 16 dots?

How many lines would be needed to make a similar pattern using 100 dots?

Summary of Lesson Block

Lesson 7B1   Children do the (test) problem LINE DESIGN (Worksheet 7B1-1).

Lesson 7B2   Comparing different methods of solution.

Lesson 7B3   At the end of the year a second test problem TENNIS COMPETITION can be administered and marked for comparison (Worksheet 7B3-1).

28 Strategies for Problem Solving

Materials required and Teacher Preparation

For Lesson 7B1, children need individual copies of the test problem (Worksheet 7B1-1), plenty of dotted circles (Worksheet 7B1-2) and spare sheets of blank paper. (Teachers will need a stapler!)

For Lesson 7B2, teachers need to read (or mark) children's solutions to their test problem and make a note of which children have used each of the different methods of solution. Suggestions for marking the test are given at the end of this lesson block.

For Lesson 7B3 (at the end of the year), children need individual copies of Worksheet 7B3-1 and plenty of paper.

SAMPLE LESSON PLAN

Lesson 7B1

1. Explain the purpose of the "test" and stress that students should explain their work to **try to convince whoever reads it that what they say is true.**

2. Hand out Worksheet 7B1-1. Read the question to the class and **make sure they understand what is meant.**

3. Ask children to do all of their working on the test sheets.
    Students who prefer to work in pairs may so do, provided
    > they work <u>very quietly</u> and
    > they put the name of anyone with whom they work on their test paper.

4. As they work, allow students to ask questions. Try to answer in such a way that not too much is given away, but some progress can be made. Teachers can record (briefly) the question and answer on the child's work at the time, if desired.

5. Collect all pieces of paper used.

6. Look at solutions before the next lesson.

    For teachers who wish to give a mark out of 10, a suggestion is to give 7 marks (appropriately distributed) for finding the number of lines for 16 dots and the remaining 3 marks for the 100 dot case. Teachers who wish to keep a more detailed record of the performance of their class, may wish to also record a score (letter or number) in some or all of the areas of

    > **working systematically**
    > **ability to generalise**
    > **efficient use of diagram**
    > **quality of explanation**
    > **method used.**

    The test at the end of the course can be assessed in a similar way, for comparison.

Lesson 7B2

1. Return test problems for duration of lesson.

2. Ask children to quickly do a LINE DESIGN problem **using 8 dots only.**

3. Explain that there are many different ways to do a problem and that **there is not necessarily a best way to do this one or any other.** This is a feature of many problem solving tasks.

4. Select children to **illustrate each of a number of different methods** at the board. To save time use 8 dots only. Methods commonly used by children are given in the Mathematical Notes.

    Most children will guess a pattern at some stage of their solution - point this out when it occurs and at that time, or later, ask children to check the pattern by using the 9 dot case, for which the answer is known.

    It is a good idea to start with Method 1, as most children - usually about 2/3 of the class - do it this way. It is also a method which is **very difficult** to use for 100 dots and the impact of Method 2 is much more dramatic if it follows Method 1. As most children use Methods 1 or 2, teachers should use their discretion as to which (if any) other methods they illustrate.

    Method 1 also leads children into looking for quick ways to sum consecutive integers - a theme which recurs in later lesson blocks.

Lesson 7B3 - AT THE END OF THE YEAR

1. The problem on Worksheet 7B3-1 can be used to measure changes in problem solving ability during the course. As the geometric model is not supplied, students find this problem considerably harder and this needs to be taken into account when making the comparison.

MATHEMATICAL NOTES

LINE DESIGN

Method 1   Counting, systematically, the "new" lines from each point.

This is the method used by most children. **Coloured pens** help make it successful.

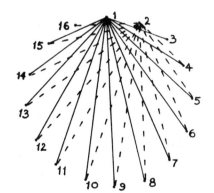

There are 13 lines from the first dot and 13 lines from the second dot, but only 12 NEW lines from third dot and only 11 NEW lines from fourth dot etc.

So the total number of lines is

$13 + 13 + 12 + 11 + 10 + \ldots + 2 + 1 + 0 + 0 = 104$

For the 100 dot pattern, the total number of lines will be

$$97 + 97 + 96 + 95 + \ldots + 2 + 1 + 0 + 0 = 4850.$$

While few children are successful with this second calculation during the test, there are many interesting ways of doing it. In the discussion, bring out the reasons why there are two thirteens and why 0 occurs twice.

For the 8 dot case, the total number of lines will be 5 + 5 + 4 + 3 + 2 + 1 + 0 + 0 = 20.

Method 2   **Multiply and then halve.**

Many children try this method but do not see why halving is necessary, so turn to Method 1.

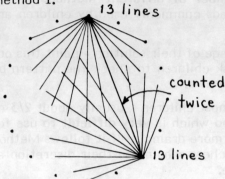

There are 13 lines at each dot and there are 16 dots, so there are 13 × 16 "ends" of lines.

As **each line has two ends,** there are $\frac{1}{2}(13 \times 16) = 104$ lines.

For 100 dots, there are $\frac{1}{2}(97 \times 100) = 4850$ lines.

For 8 dots, there are $\frac{1}{2}(5 \times 8) = 20$ lines.

Here is a solution from a Year 8 student:

"Firstly you take away three dots from your original amount of dots, then you multiply this number by the original amount. The figure you now receive you divide by two. And you now have the amount of lines.

$$x = (n - 3) \times n \div 2 \qquad x = \text{amount of lines} \qquad n = \text{amount of dots.}$$

You take the three dots away because you don't count yourself or your neighbours. Then you multiply the original from step 1 to get the number of points coming out from each dot. As you have done this you will have counted every line twice so you divide by two."

Method 3   **Try small cases first and GUESS at a pattern.**

Number of dots	Design	Number of lines	Difference
3	...	0	
			2
4	✦	2	
			3
5	✶	5	
			4
6	✹	9	
			5
7	✺	14	
			predict 6
8		Check by drawing or other means	

This is essentially an inductive method of solution. Some teachers use it, but very few year 7 or 8 children do. If this method is demonstrated, emphasise that the pattern needs to be checked by drawing or other means on other small cases (e.g. 8 dots) before it can be confidently used to predict answers. As well as checking on other cases, it is useful to think about why the pattern might hold: when a new dot is added, how many more lines come into the design?

Method 4   Counting from a different angle.

This method seems to occur only occasionally, but it gives a different insight to the problem.

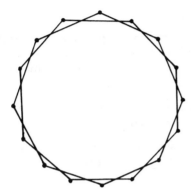

Figure 1

Figure 2

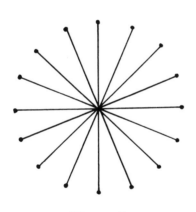

Figure 3

There are   16 lines skipping **one** point (figure 1),
16 lines skipping **two** points (figure 2),
16 lines skipping **three** points,
16 lines skipping **four** points,
16 lines skipping **five** points,
and   16 lines skipping **six** points.

However, there are only 8 lines skipping **seven** points (figure 3) as these join opposite points.

Skipping more than seven points gives a line already counted, for example skipping 8 points is the same as skipping 6 points.

So the total number of lines is
6 × 16 + ½ × 16 = 104.

Note that this method gives a **slightly different pattern for odd and even numbers of dots.** For example, with 15 dots there are 15 lines skipping each of 1, 2, 3, 4, 5 and 6 points. There is no set of lines joining opposite points. So the total number of lines is 6 × 15 = 90. For 100 dots, the total number of lines is 48 × 100 + ½ × 100 = 4850.

TENNIS COMPETITION

This problem has been chosen as the second test problem because the methods for LINE DESIGN can be used to solve it and the same problem solving strategies are useful. However it is more difficult as children have to create their own geometric interpretation. If each team is designated by a point and each game is designated by a line joining the opposing teams, in the "line design" so created, each dot is joined to all but one of the other dots. Using Method 2 (above) there will be ½ × 16 × 14 games with 8 teams and ½ × 100 × 98 games with 50 teams.

Strategies for Problem Solving　　　　　　　　　　　Worksheet 7B1-1

Name: _____　　　　Partner's Name: _____

LINE DESIGN

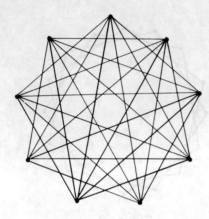

This pattern was made by placing nine dots in a circle and joining each dot to every other dot, except its nearest neighbours. The pattern needed 27 lines.

How many lines would be needed to make a similar pattern using 16 dots?

How many lines would be needed to make a similar pattern using 100 dots?

Strategies for Problem Solving

Worksheet 7B1-2

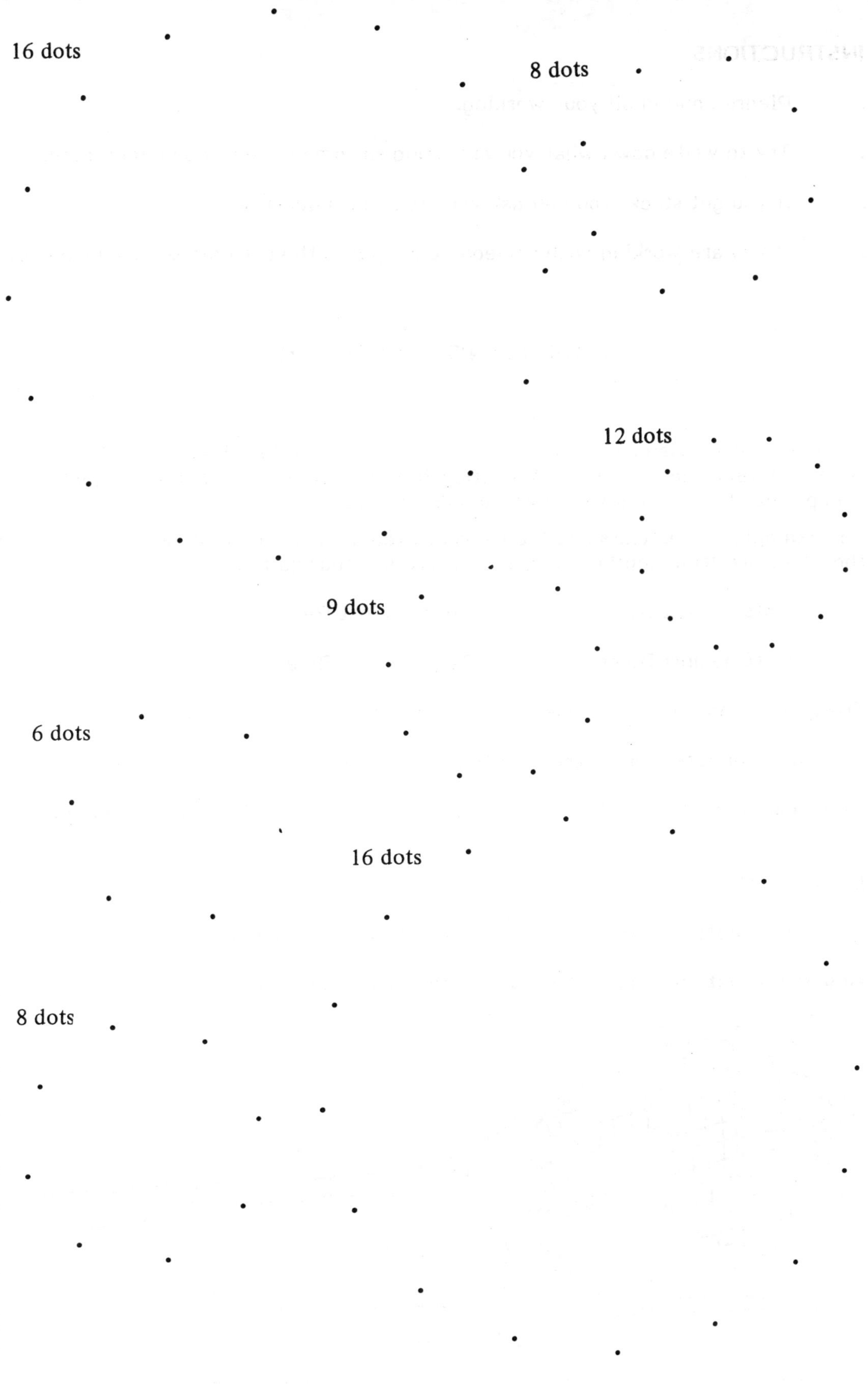

16 dots

8 dots

12 dots

9 dots

6 dots

16 dots

8 dots

Strategies for Problem Solving   Worksheet 7B3-1

INSTRUCTIONS

- Please hand in all your working.

- Try to write down what you are doing so someone else can understand.

- If you get stuck, you can ask your teacher a question.

- If you are working with someone else, put both your names on your paper.

THE TENNIS COMPETITION

A number of tennis clubs are having a competition and each club is entering two teams. Every team in the competition has to play every other team just once, except that teams from the same club do not play each other.

For example, if the Cats and the Dogs are teams from one club and the Reds and the Blues are from another club, then there are four games:

    Cats against Reds   Dogs against Reds

    Cats against Blues   Dogs against Blues

The Cats do not play the Dogs and the Reds do not play the Blues.

How many matches will there be altogether if 8 clubs enter the competition?

How many matches will there be altogether if 50 clubs enter the competition?

IF YOU FINISH EARLY:

How many matches would there be if each club entered 3 teams?

How many matches would there be if each club entered 4 teams?

Lesson Block 7C

PATTERN SPOTTING

Theme   An introduction to pattern spotting.

Aims   To meet some elementary techniques for **spotting patterns** in sequences of numbers.

To learn how to **extract from a problem** the information needed to find **a useful pattern.**

Mathematical Activity

TRIANGLE DESIGN

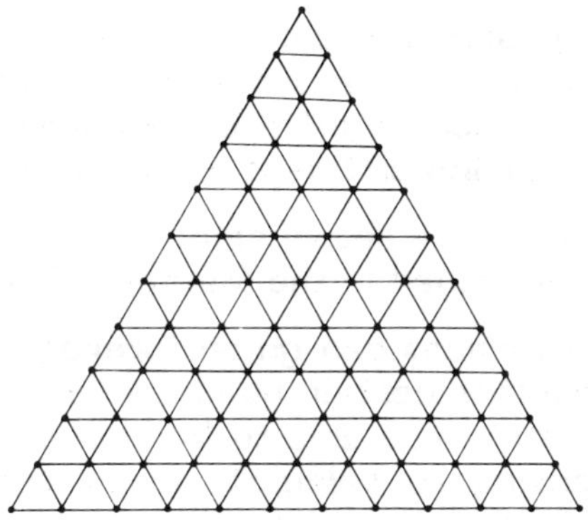

What patterns can you see here?

Summary of Lesson Block

Lesson 7C1   **Spotting patterns** in number sequences and using these to **extend** the sequence. (Approximately two periods.)

Lesson 7C2   **Extracting number patterns** from the TRIANGLE DESIGN by **trying plenty of examples, guessing** a pattern, using the guess to **predict** a particular case and **checking** the predicted answer.

Materials Required and Teacher Preparation

Copies of Worksheets 7C1-1, 7C1-2 and 7C2-1. Teachers should examine the sequences used on the worksheets and select examples of the appropriate difficulty. Do not ask children to do all the examples on Worksheet 7C1-1.

36 Strategies for Problem Solving

SAMPLE LESSON PLAN

Lesson 7C1

1. **Refer back** to SQUARES ON A CHESSBOARD and LINE DESIGN and remind children of some of the patterns they used to find solutions. **Discuss** briefly how finding these patterns helped them to find answers for the large cases – without them they would still be counting!

2. This lesson is about **pattern spotting.** **Illustrate** some basic ways of spotting patterns by:

    . looking at how successive numbers in a string are connected to one another (in particular by examining **differences** between successive numbers and **ratios** of successive numbers).

    . looking at how each number in a sequence is connected to its **"position number".**

    The Mathematical Notes suggest how these ideas can be explained.

3. Children do examples from Worksheet 7C1-1.

4. **Discuss** answers, taking particular care at times to **check** that the pattern found works for <u>all</u> given terms of the sequence. **Encourage** discussion about any **different patterns which lead to different sequences, different ways of finding the same sequence** and **finding different patterns in the same sequence.**

5. Reinforce these ideas using the sequence of square numbers and the example 0, 1, 10, 11, 100, ... as explained in the Mathematical Notes.

6. Explain that number sequences can **go on forever** and that it is often important to be able to decide whether certain numbers can be part of a given sequence or not. It is also very helpful to be able to find numbers occurring a long way down a sequence **without having to calculate every term.** (This is where **position numbers** are so useful.) Discuss the meaning of the questions on Worksheet 7C1-2 with the children and get them to complete them individually or in pairs. Allow plenty of time for discussion of answers to question 2.

Lesson 7C2

1. (Teacher-led discussion based on Worksheet 7C2-1)

    What kind of <u>number</u> patterns can we find in the <u>geometric</u> TRIANGLE DESIGN?

    Use this to **illustrate** the importance of
    . trying plenty of examples
    . working systematically
    . keeping a record (tables are particularly useful here).

    After **guessing** a pattern, it is important to **check** that it works by
    . predicting an answer, using the pattern
    . using a completely different method (eg. counting)
    . seeing if the two answers agree.

    Emphasise these points with a chalkboard summary.

2. Ask children to work individually, or in pairs, on a TRIANGLE DESIGN problem of your (or their own) choice. A report on these discoveries should be completed for homework.

Lesson Block 7C   37

MATHEMATICAL NOTES

POSITION NUMBER, DIFFERENCES AND RATIOS

Look at the sequence 3, 6, 9, 12, , , , .

You can think of this pattern in two ways—

3, 6, 9, 12, 15, 18, 21
  +3 +3 +3 +3 +3 +3

ie. adding 3 to each number to get the next one (looking at **differences** between successive numbers).

or as

3, 6, 9, 12, 15, 18, 21
3x1 3x2 3x3 3x4 3x5 3x6 3x7

ie. each number is 3 times its **position number.**

What pattern can you see in the sequence 3, 9, 27, 81, 243, 729, , , , ?

Some people see

3, 9, 27, 81, 243, 729,
  x3 x3 x3 x3 x3 x3

ie. multiplying each number by 3 to get the next one. (Looking at the **ratio** of successive numbers)

Other people see

3 = 3,	9 = 3 x 3,	27 = 3 x 3 x 3,	81 = 3 x 3 x 3 x 3
1 lot of 3	2 lots of 3	3 lots of 3	4 lots of 3.

Here the **position number** tells how many threes to multiply together.

DIFFERENT PATTERNS: SAME OR DIFFERENT NUMBERS?

Sometimes **different patterns give different numbers.** If two different patterns both fit all the numbers given, you need some other information to decide which is best.

What pattern can you see here?

$$0, 1, 10, 11, 100, 101, , , ,$$

Most children say that the next numbers are 1000, 1001, 10000 but teachers and others who know about binary numbers often see a different pattern and think the next numbers are 110, 111, 1000. There's no way of deciding who's right until we have other information.

38 Strategies for Problem Solving

Notice that there are often **different ways of finding the same numbers!** Did you find this pattern?

$$1, \quad 4, \quad 9, \quad 16, \quad 25, \quad 36, \quad 49$$
$$\ \vee\ \ \vee\ \ \vee\ \ \vee\ \ \vee\ \ \vee$$
$$\ +3\ \ +5\ \ +7\ \ +9\ \ +11\ \ +13$$

or this one

$$1, \quad 4, \quad 9, \quad 16, \quad 25, \quad 36, \quad 49$$
$$\downarrow \quad \downarrow \quad \downarrow \quad \downarrow \quad \downarrow \quad \downarrow \quad \downarrow$$
$$1\times1 \ \ 2\times2 \ \ 3\times3 \ \ 4\times4 \ \ 5\times5 \ \ 6\times6 \ \ 7\times7$$

Both patterns are right and they both produce the same numbers.

FINDING NUMBER PATTERNS

Suggested solutions to Worksheet 7C1-1

A.	Add 2	AS	K.	Add 1, 2, 3, ...	AS
B.	Add 3	AS	L.	Multiply by 3, 2, 3, 2,...	MD
C.	Subtract 6	AS	M.	Squares	P
D.	Subtract 10	AS	N.	Cubes	P
E.	Double	MD	O.	Add previous two terms	AS
F.	Triple	MD	P.	Subtract 2	AS
G.	Halve	MD	Q.	Subtract 3	AS
H.	Divide by 5	MD	R.	Add 2, 3, 4, ...	AS
I.	Multiply by 0.9	MD	S.	Multiply by -2	MD
J.	Omit multiples of 3	O	T.	Binary numbers	P

Solutions to Worksheet 7C1-2, Question 2

10TH NUMBER	100TH NUMBER
-9	-99
100	10000
$10/11$	$100/101$
$3\ 1/4$	$25\ 3/4$
110	10100

Strategies for Problem Solving   Worksheet 7C1-1

FINDING NUMBER PATTERNS

What numbers do you think go in the spaces?

As you fill in each number pattern, use the box at the right to show how you found your pattern – if you looked at what you have to **add** or **subtract** write AS; if you **multiplied** or **divided** each number to get the next one, write MD; if you found a pattern using the **position number** write P; and if you used none of these write 0.

For example, in the first box you could write AS or P.

A.	2, 4, 6, 8, , , ,	☐	K.	5, 6, 8, 11, 15, , , ,	☐
B.	4, 7, 10, 13, , , ,	☐	L.	2, 6, 12, 36, 72, , ,	☐
C.	100, 94, 88, 82, , , ,	☐	M.	1, 4, 9, 16, , , ,	☐
D.	30, 20, 10, 0, , , ,	☐	N.	1, 8, 27, 64, , , ,	☐
E.	2, 4, 8, 16, , , ,	☐	O.	1, 3, 4, 7, 11, , , ,	☐
F.	1, 3, 9, 27, , , ,	☐	P.	-2, -4, -6, -8, -10, , , ,	☐
G.	10, 5, 2½, 1¼, , , ,	☐	Q.	-2, -5, -8, -11, -14, , , ,	☐
H.	20, 4, 0.8, 0.16, , , ,	☐	R.	-7, -5, -2, 2, 7, , , ,	☐
I.	1, 0.9, 0.81, 0.729, , , ,	☐	S.	2, -4, 8, -16, 32, , , ,	☐
J.	1, 2, 4, 5, 7, 8, 10, , , ,	☐	T.	0, 1, 10, 11, 100, , , ,	☐

Strategies for Problem Solving

Worksheet 7C1-2

EXTENDING NUMBER PATTERNS

1. Complete the table by looking at the number patterns on the left and putting a circle around any numbers on the right that belong to the pattern. (Remember that the number patterns "go on forever", so the numbers you circle do not have to be one of the next three numbers in the pattern.)

PATTERN	NUMBERS TO CHECK
5, 10, 15, 20, , , ,	73, 104, 135
1, -2, 3, -4, 5, , , ,	12, -31, 49
1, 4, 16, 64, , , ,	128, 256, 100 001
1/2, 3/4, 5/6, 7/8, , , ,	99/100, 100/101, $1\tfrac{1}{2}$

2. Complete the table by writing down the 10th number in each of the number patterns. In the last column, either write down the 100th number in the pattern <u>or</u> try to say how to find it.

NUMBER PATTERN	10TH NUMBER	100TH NUMBER
0, -1, -2, -3, ...	-9	-99
1, 4, 9, 16, ...		
1/2, 2/3, 3/4, 4/5, ...		
1, $1\tfrac{1}{4}$, $1\tfrac{1}{2}$, $1\tfrac{3}{4}$, 2, ...		
2, 6, 12, 20, 30, ...		

Strategies for Problem Solving   Worksheet 7C2-1

THE TRIANGLE DESIGN

Carol and I made this design by sticking matches onto a piece of cardboard. I started at the top (level 1) with 3 matches

then Carol added 6 more matches at level 2

and so on.

1. **Fill in** the gaps.

Level	1	2	3	4	5
Number of matches	3	6			

2. **Guess** the number of matches at level 10.

3. **Check** your guess by counting the number of matches at level 10.

   Was your guess correct?

4. Which level would use 39 matches?

5. Match boxes contain 47 matches. Does any level use **exactly one matchbox?**

6. Nicholas decided to make a similar design with 15 levels.

   How many matches would Nicholas need altogether?

7. Vivienne decided that she would use some coloured triangle stickers to make a design like ours.

   Think of some questions for which Vivienne might want to know the answers.

8. With the help of your teacher, choose one of Vivienne's Triangle questions and find the answer.

Lesson Block 7D

UNDERSTANDING THE QUESTION

Theme Read the question – really understand it.

Aims To encourage students to think carefully about what a problem requires them to do and about the information they are given.

To alert students to three major traps:

(i) thinking you want something you don't (Lesson 7D1)
(ii) not noticing some information in the question (Lessons 7D1 and 7D2)
(iii) reading into the question something that is not there (Lesson 7D3).

Mathematical Activity

SHORT PUZZLES

This theme is illustrated by a series of short questions selected from Worksheets 7D1-1 and 7D3-1. They include puzzles, brain teasers and tricks, but they illustrate strategies that are important in serious work.

Summary of Lesson Block

Lesson 7D1 Working out what you **WANT** and what you **KNOW** with ST IVES, HOLE NUMBERS and other problems from Worksheet 7D1-1.

Lesson 7D2 The ideas of Lesson 7D1 are reinforced by continuing work from Worksheet 7D1-1.

Lesson 7D3 Beware of making "Hidden Assumptions". This is introduced by a discussion of SPLAT ! and other questions from Worksheet 7D3-1.

Materials Required and Teacher Preparation

For Lessons 7D1 and 7D2, Worksheet 7D1-1 is required. Worksheet 7D2-1 is used to introduce Lesson 7D2 and revise points made previously. For Lesson 7D3, students need copies of Worksheet 7D3-1. A model, for example 6 matches and some "BluTak", is useful to illustrate the solution to TRI-FOUR.

Teachers can watch "normal" mathematics lessons for good examples of students getting stuck because they are unclear about what they want, or what information they are given. These examples will be useful in Lessons 7D1 and 7D2.

SAMPLE LESSON PLAN

Lesson 7D1

1. A **very common reason for being stuck is not understanding the question completely.** Explain that in "Helpful Hints for Problem Solvers" (Lesson Block 7A) students were told to really read the question. These lessons are about what that means and how they might do it.

    Briefly give an example from a recent "normal" mathematics lesson of someone being stuck because they hadn't read the question carefully.

2. **Write on Chalkboard**

    Statements of problems tell you two things

    (i) exactly what you WANT to find - i.e. what you WANT;
    (ii) some information to use in finding it - i.e. what you KNOW.

    Offer the hint that it is frequently **a good idea to write these down in your own words** before you start a problem. Mention, perhaps, that you'll also have **to use things you KNOW** that aren't given directly in the question and **what you WANT may be modified** as you go along.

3. **WHAT DO I WANT?**

    Instruct class to listen carefully to the following question and to silently write down in their books

    (i) exactly what they WANT to find,
    (ii) the important things they are told (what they KNOW),
    (iii) the answer to the question.

    Insist that there be no calling out or comparing of answers. Read ST IVES twice, then ask for WANT and KNOW answers from selected students. When these have been clarified, make sure all students know the answer to the puzzle.

    Repeat the procedure above with HOLE NUMBERS.
    Point out that although these are tricks, **the importance of clarifying what you WANT is just as great in "ordinary" maths.**

4. **WHAT DO I KNOW ?**

    Tell students the trick questions are now over, but to stay alert.

    Do RIDING STABLE with the class. The problem is very easy, so use it for a discussion on how you can organize the information so that it's easy to use.

    One good way is to display the data in a table.

Breed	Size		
	Stallion	Mare	Pony
Arabian	✓	✓	✓
Palomino	✓	✓	✓
Pinto	✓		✓
Apaloosa	✓	✓	

5. If time allows, follow up with COLOUR MATES or MAKEUP where a table is also useful for organizing the information.

44 Strategies for Problem Solving

Lesson 7D2

1. Hand out the individual copies of Worksheet 7D2-1 "Read the Question - Really Understand It". Use this as a basis for revision of the two major points of the last lesson. The third point (Beware of Hidden Assumptions) will be covered in the next lesson.

2. Suggest that students finish at least one of MAKE UP or COLOUR MATES to practise using a table for organizing information. Discuss solutions on the board.

3. Allow students to work on problems from Worksheet 7D1-1. Suggest they start each problem by **thinking about (and writing down) what they KNOW and WANT**. Discuss solutions with the class, pointing out where the difficulties usually lie.

4. Finish the lesson by referring back to the summary on "Read the Question - Really Understand it".

Lesson 7D3

1. **WHAT DO I KNOW ? - DON'T ASSUME THINGS THAT AREN'T THERE.**

    Explain how people sometimes read into a question all sorts of constraints that aren't there. **Getting out of these mental straight-jackets is an important part of creative thinking.**

    Introduce SPLAT ! and ask each child to jot down three or four explanations and then see if you can combine them to get about ten on the board. To stop ideas spreading, ensure writing is done quietly. **Emphasise how easy it is to make assumptions unconsciously.**

2. Let children try NINE DOTS. As children try the question, give hints which gradually release them from the hidden assumptions they have unconsciously made. Discuss why this puzzle is hard - most people find it hard because there is a tendency to read into the question the constraint that straight lines must stay within the boundary of the square.
(Ask anyone who has seen it before to use only three straight lines by upsetting another hidden assumption !)

3. Allow children to work on TRI-FOUR and the SNEAKY RIDDLES. In class discussion of the solutions, **point out the hidden assumptions that hinder progress.** Get children to write these down as part of their solutions.

MATHEMATICAL NOTES

ST IVES

Probably only one person - the others were heading away from St. Ives.

HOLE NUMBERS

Holes contain no dirt.

NESSY

This problem illustrates the importance of understanding exactly what a question means before trying to answer it. The only way the sentence makes sense is if the total length is the sum of 20 metres and half the length. i.e. if the monster is 40 metres long.

RIDING STABLE

Organising information by drawing up a table (see Sample Lesson Plan).

TAXI TALK

Get **all** the information out of the question. The driver heard where she wanted to go.

MONDAYITIS

Today is Monday.

TRAIN TUNNEL

Making a model or drawing a diagram are both important ways of understanding a question. Answer: 2 minutes.

COLOUR MATES

	BOYS		
GIRLS	red	green	blue
red	X	X	✓
green	✓	X	X
blue	X	✓	X

The solid entries are obtained from information given in the question. The dotted entries have been logically deduced from this information. The girl in red was dancing with the boy in blue.

MAKE-UP

	Nails	Make-up	Reading	Hair
Myra	X	✓	X	X
Maud	X	X	✓	X
Mary	X	X	X	✓
Mona	✓	X	X	X

The solid entries are obtained from information given in the question. The dotted entries have been deduced from this information. Myra is putting on make-up, Mary is doing her hair, Mona is doing her nails. Maud is their mother.

SPLAT!

Some of the following explanations will expose some unwarranted assumptions.

1. Man hits tree, stops, skis continue on.
2. Skis sent down alone.
3. One ski sent down twice.
4. One man goes on one ski twice.

46 *Strategies for Problem Solving*

5. Man skis to tree, detaches skis, walks around tree, refastens skis in same tracks.
6. Two men go down, outside marks erased.
7. Short tree - man goes over it.
8. Tree placed after tracks are made.
9. Pliant tree bends over as skier straddles it.
10. Man goes to tree, lifted over, continues on.
11. Hill is part of narrow valley - man goes down on one ski, up the opposite slope and back past tree on return trip.
12. Gentle slope - man lifts one ski to pass tree and replaces it as he passes to show a continuous track.
13. Man goes through tree!!

NINE DOTS

Once a person recognizes that nothing in the problem imposes staying within the array, the solution is usually forthcoming.

By overcoming other hidden assumptions (e.g. the points are mathematical ones having no size or the paper can't be folded) fewer than four lines are needed.

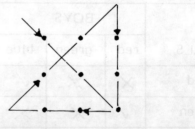

TRI-FOUR

Make a tetrahedron, thus overthrowing the hidden assumption that the triangles all have to be in the one plane.

SNEAKY RIDDLES

Watch for unjustified assumptions.
1. It was daytime.
2. ISLAND.
3. He was blind.
4. It was a tin of coffee powder.
5. The father is bald.

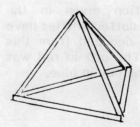

[These riddles are from **"aha! Insight"** by Martin Gardner, Scientific American, 1978, New York.]

Strategies for Problem Solving   Worksheet 7D1-1

ST IVES

As I was going to St Ives,
I met a man with seven wives,
Every wife had seven sacks,
Every sack had seven cats,
Every cat had seven kits.
Kits, Cats, Sacks and Wives,
How many were going to St Ives?

HOLE NUMBERS

How much dirt is there in a hole measuring 1.01 metres by 2.02 metres by 3.03 metres?

NESSY

The length of the Loch Ness monster is 20 metres long and half its own length, but it isn't 30 metres long. How many metres long is it?

RIDING STABLE

A riding stable owner has the following kinds of horses and ponies: Arabian mare, Apaloosa stallion, Palomino pony, Arabian pony, Palomino mare, Pinto stallion, Apaloosa mare, Palomino stallion, Pinto pony and Arabian stallion. If she wants to have mounts of every size and breed combination, what types of horses or ponies should she obtain?

TAXI TALK

One day a lady caught a taxi. On the way to her destination the lady talked so much that the driver became quite annoyed. Finally the driver said: "I'm sorry madam but I can't hear a word you're saying. I'm as deaf as a post and my hearing aid hasn't worked all day." When she heard this the lady stopped talking. But after she left the cab she suddenly realised that the driver had lied to her.
How did she know?

MONDAYITIS

When the day after tomorrow is yesterday, today will be as far from Monday as today was from Monday when the day before yesterday was tomorrow. What day is it now?

TRAIN TUNNEL

A freight train 1km long is travelling at 60 km per hour when it enters a tunnel also 1km long. How long does it take the train to go through the tunnel?

COLOUR MATES

A taxi picked up three young couples and took them to a discotheque. One girl was dressed in red, one in green, and one in blue. The boys wore outfits of the same three colours. When all the three couples were dancing, the boy in red danced close to the girl in green and pointed out to her that each one of them was dancing with a partner dressed in a different colour.
What colour is the partner of the girl in red wearing?

MAKE-UP

Three girls are getting ready for the disco while their mother is reading. One girl is doing her nails, one is doing her hair and one is putting on her make-up.

1. Myra is not doing her nails and isn't reading.
2. Maud is not putting on make-up or doing her nails.
3. Mary is not reading or doing her nails.
4. Mona is not reading and is not putting on make-up.
5. Myra is not doing her hair.

What is each girl doing?

Some of these questions are from **"aha! Insight"** by Martin Gardner, Scientific American, 1978, New York.

Strategies for Problem Solving   Worksheet 7D2-1

READ THE QUESTION — REALLY UNDERSTAND IT!

Remember this hint from "Helpful Hints for Problem Solvers"? It's an important one because not understanding the question is a common reason for being stuck.

Sometimes you get stuck because you aren't clear about what you want to find...

...sometimes because you haven't noticed some vital information...

...and sometimes you unconsciously assume something wrong.

So, when you start a question, read it two or three times carefully.

You're not ready to move on until you can explain the question in your own words without looking at it.

Whenever you get stuck, pause and ask yourself:—

What do I really WANT to find?
What do I KNOW that will help?

Illustrations by Cathy Hefford

Strategies for Problem Solving   Worksheet 7D3-1

BEWARE OF HIDDEN ASSUMPTIONS

SPLAT!

Coming down a hill covered with snow are two ski tracks. One track goes around a tree on one side. The other track goes around the tree on the opposite side. Give 10 different physical explanations to fit these facts.

NINE DOTS

Without lifting your pencil from the paper, draw four straight line segments through the dots. (You cannot go back over a line.)

• • •
• • •
• • •

TRI-FOUR

With six sticks of equal length how can you form four equilateral (all sides equal) triangles without breaking or cutting the sticks?

SNEAKY RIDDLES

When the music stopped the six friends at the disco returned to their table and amused themselves by asking each other riddles.

How many can you get?

The boy in red asked his first:
"Last week I turned off the light in my bedroom and managed to get to bed before the room was dark. If the bed was 3 metres from the light switch, how did I do it?"

The boy in green said:
"What common word starts with "IS" ends with "ND", and has "LA" in the middle?"

The girl in red said:
"One night my uncle was reading an exciting book when his wife turned out the light. Even though the room was pitch dark he went right on reading. How could he do that?"

The girl in green said:
"This morning one of my earrings fell into my coffee. Even though my cup was full the ring didn't get wet. How come?"

The girl in blue asked the last riddle.
"Yesterday, my father was caught in the rain without a hat or umbrella. There was nothing over his head and his clothes got soaked. But not a hair on his head got wet. Why was that?"

[The sneaky riddles are from **"aha! Insight"** by Martin Gardner, Scientific American, 1978, New York.]

Lesson Block 7E

USING PATTERNS

Theme How patterns can be used to solve problems.

Aims To remind children of the importance of **trying plenty of examples**, **working systematically** and **keeping a record** when trying to guess a pattern.

To emphasise the importance of **checking** and to illustrate how to check.

To illustrate how, in order to find useful patterns, it is often necessary to produce simpler cases by varying the numbers in the original problem.

Mathematical Activity

HANDSHAKES

At the beginning of a meeting of The Society For The Preservation of Good Manners, everyone shakes hands with everyone else. One night 30 people came to a meeting.

How many handshakes were there all together?

Summary of Lesson Block

Lesson 7E1 Children do problems from Worksheets 7E1-1 and 7E1-2. Different methods of solution are discussed.

Lesson 7E2 Children continue working on problems and write up a solution to one problem of their choice.

Lesson 7E3 Procedures for **checking** are revised and children do REGIONS OF A CIRCLE from Worksheet 7E3-1.

Materials Required and Teacher Preparation

Before reading the Mathematical Notes, teachers should attempt all of the problems, **without using algebra**. The Mathematical Notes discuss patterns commonly found and used by children.

The Chalkboard Summary should be written up before, or at the beginning of, Lesson 7E1.

As well as individual copies of Worksheets 7E1-1, 7E1-2 and 7E3-1, children will require **scrap paper**, some **headless matches** and, perhaps, some **grid paper** for drawing match patterns and making up tables.

SAMPLE LESSON PLAN

Lesson 7E1

1. Explain to children that this lesson block is about **using number patterns** to help to solve problems. Discuss what is required in order to be able to **guess** a number pattern and how it can be **checked** - refer to the Chalkboard Summary.

Chalkboard Summary

PATTERNS HELP YOU SOLVE PROBLEMS

To **guess** a pattern, you will need to

- try plenty of examples
- work systematically
- keep a record of what you have done

(tables are often a good way to keep a record).

After **guessing** a pattern, you will need to **check** that it works.

To **check your guess,** try some examples you can work out two ways

- using your pattern

  and

- by a completely different method

(for example, drawing a picture and counting).

---

2. Hand out Worksheet 7E1-1. Make sure children understand the MATCH PATTERN problem, then allow children some time to work on the problem individually or in pairs.

3. When some children have solved the problem, discuss their solutions and relate them to the Hint on the Worksheet and the Chalkboard Summary. Solutions along the lines of Methods 1 and 2 in the Mathematical Notes are neater and offer more insight than Method 3. However, it is worth illustrating Method 3 as well and explaining its great advantage - it is a good way of getting started if you are stuck.

4. Ask children to work on HANDSHAKES and (later) the problems on Worksheet 7E1-2. When appropriate, discuss different children's solutions and introduce new problems. (While children are working on the problems, the teacher can select children who have different methods of solution, to illustrate their solution at the board - this gives children an opportunity to practise **explaining what they have done in a way that someone else can understand.**)

Lesson 7E2

1. Continue working on problems as in part 4 of Lesson 7E1. Demonstrate how the pattern strategy can be used for each problem.

2. Ask childen to **write up** one problem of their choice under the headings "What I discovered" and "What problem solving idea I found helpful". They should summarise, rather than repeat, what has been done. (This part may need to be completed as homework.)

52 Strategies for Problem Solving

Lesson 7E3

1. Remind children that spotting a pattern is really making a **guess** and that guesses can't be trusted until they have been **checked**. (Older students should remain sceptical until they have a reason which explains why the pattern holds, but this can't be pushed too far at this level.)

2. Revise the procedure for **checking** using a problem done earlier, such as ODDER AND ODDER.

    PROCEDURE FOR CHECKING

    After **guessing** a pattern, it is important to **check** that it works by

    - predicting an answer, using the pattern,

      (e.g. the sum of the first 5 odd numbers is 5x5);

    - finding the answer by some completely different method,

      (e.g. $1 + 3 + 5 + 7 + 9 = 25$);

    - comparing the actual and the predicted answers -

      if they are different, the **guess** needs to be revised, while if they are the same (as in the above example, where $25 = 5 \times 5$) it doesn't **prove** that the pattern is correct, but only provides supporting evidence

    and, wherever possible,

    - finding a reason to **explain why** the pattern holds.

    Children are often surprised when this process is explicitly explained to them - they frequently think that checking refers only to doing the arithmetic calculation again to check that no mistakes have occurred.

3. Hand out Worksheet 7E3-1 and ask children to work individually or in pairs on REGIONS OF A CIRCLE.

4. Discuss children's results for REGIONS OF A CIRCLE and how this problem illustrates the importance of **checking**.

5. **Write up** REGIONS OF A CIRCLE, including some comment on the importance of checking. (One teacher suggested to her class that they use the heading of "A Cautionary Tale" for this problem.)

MATHEMATICAL NOTES

MATCH PATTERNS

Method 1

A track 90 cm long requires 91 matches. One way to see the answer is as twice the width (for the top and bottom) plus the number of vertical matches. The width must be measured in matches!

Method 2

A nicer method is to start with 1 match and add 3 matches the required number of times. This immediately gives $3n + 1$, where n is the number of squares.

Method 3

This method depends on trying simpler cases (i.e. with a shorter track), recording the result and hence finding the pattern. It is not as neat as the previous two methods and offers less insight into the situation, but it is an easy way to get started, which is great if you are stuck!

Diagram	I	□	⊏⊐	⊏⊓⊐	...
Length (in cm)	0	3	6	9	...
Number of matches	1	4	7	10	...

HANDSHAKES

The answer is 435 handshakes for 30 people. A few children recognise that this is essentially the same problem as LINE DESIGN (Lesson Block 7B). For different methods of solution, see the Mathematical Notes in that section.

This is a good problem to use **guess and check** - when children think they have found a pattern they can use it on **an easier example** (e.g. 3 or 4 people) and **check** their answer by actually shaking hands and counting the number of handshakes. (This is a particularly good teaching strategy to adopt when children **guess the wrong pattern** and say that for 30 people there will be 29 x 30 handshakes.)

PAPER FOLDING

Most children try to fold the paper 10 times - only to discover that it is impossible. Very few keep a record of the number of creases as they fold.

After making a table such as

Number of folds	1	2	3	4	...
Number of creases	1	3	7	15	...
Number of thicknesses	2	4	8	16	...

children notice patterns like

(i) the difference between the number of creases is 2, 4, 8 ...,

(ii) the number of creases at each stage is one more than twice the number of creases immediately before.

Few children recognise that the numbers of creases are one less than the powers of 2.

Adding a record of the number of thicknesses at each stage to the table provides more patterns to work with and leads to the easiest explanation of why the patterns are correct. Each new fold naturally adds a new crease **to each thickness.**

54 *Strategies in Problem Solving*

ODDER AND ODDER

A discussion on which number **is** the 50th odd number can be profitable. It can also be used to introduce the idea of **try an easier example.** (Here, ask: "What is the 5th odd number ?")

Common methods children use to find the sum are:

(i)
$$1 + 3 + 5 + 7 + 9 = 25,$$
so $$11 + 13 + 15 + 17 + 19 = 25 + 50,$$
and $$21 + 23 + 25 + 27 + 29 = 25 + 50 + 50,$$
etc.

(ii) Looking at simpler cases gives

$$1 = 1$$
$$1 + 3 = 4$$
$$1 + 3 + 5 = 9$$
$$1 + 3 + 5 + 7 = 16$$
$$1 + 3 + 5 + 7 + 9 = 25$$

leading to **"the sum of the first n odd numbers is n^2".**

(iii) $1 + 99 = 100$, $3 + 97 = 100$, $5 + 95 = 100$, ... , $49 + 51 = 100$.

So the sum of the first 50 odd numbers is 25 lots of $100 = 2\,500$.

A very nice geometric illustration of the fact that the sum of the first n odd numbers is n^2 is shown below.

```
                                              + + + +
                          x x x               x x x +
             o o          o o x               o o x +
   x         x o          x o x               x o x +

 1 = 1,    1+3 = 2²,    1+3+5 = 3²,       1+3+5+7 = 4²
```

When children have seen both methods (ii) and (iii), a good **"What if...?"** to ask is

What is the sum of the first 100 odd numbers?

You can **check** that both methods give the same answer, namely
$$100^2 = 50 \times (1 + 199).$$

REGIONS OF A CIRCLE

This problem illustrates beautifully the dangers of **guessing** a pattern from the first few terms of a sequence of numbers. Naturally, everyone guesses that there are 32 regions for 6 dots when in fact there are only 30 when the dots are evenly spaced (and at most 31 otherwise).

Unfortunately, the real underlying pattern is too difficult to explain to children.

However, for the benefit of teachers, the **maximum** number of regions for n dots is

$$1 + {}^nC_2 + {}^nC_4 = (n^4 - 6n^3 + 23n^2 - 18n + 24)/24,$$

(where, for n=1 both ${}^nC_2$ and ${}^nC_4$ are omitted and, for n = 2 or 3, ${}^nC_4$ is omitted).

This can be calculated by summing the first five terms in Pascal's triangle thus:

```
                    TRIANGLE                                    SUM

                       1                                         1
                    1     1                                      2
                 1     2     1                                   4
              1     3     3     1                                8
           1     4     6     4     1                            16
        1     5    10    10     5  /  1                         31
     1     6    15    20    15  /  6     1                      57
  1     7    21    35    35  / 21    7     1                    99
```

When n is odd, the number of regions for regularly spaced dots is equal to the maximum, but for even n there will always be fewer regions (the exact number not being known for all cases).

Strategies for Problem Solving   Worksheet 7E1-1

MATCH PATTERNS

A match is 3 cm long. It takes 16 matches to make a track 15 cm long and 3 cm wide, like this:

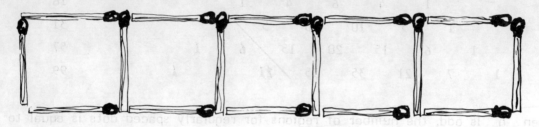

How many matches does it take to make a similar track 90 cm long and 3 cm wide?

Hint: You probably don't have enough matches to make a 90 cm track. So try to **find a pattern to help you solve this problem.**

To do this, **try some examples** - i.e. make some smaller tracks and **record** your results.

Now **guess** what the pattern might be. Then **check** your guess on some examples you can work out two ways - by drawing and counting **and** by using your pattern.

You can now **use your pattern** to solve the problem.

HANDSHAKES

At the beginning of a meeting of The Society For The Preservation Of Good Manners, everyone shakes hands with everyone else. One night 30 people came to a meeting.

How many handshakes were there altogether?

Hint: Use **something to help you** - perhaps some friends for shaking hands, or a picture or a diagram.

If you are **stuck, try some examples** - in this case, try some meetings with a small number of people first.

Guess a pattern and **check** it - by counting handshakes **and** by using your pattern.

Now **use your pattern** to solve the problem.

Strategies for Problem Solving   Worksheet 7E1-2

PAPER FOLDING

Take a strip of paper about 30 cm long and 2 cm wide. Fold it in half, then in half again. When you unfold it you will see that 3 creases have been made.

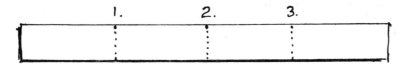

If you folded the strip in half **ten times**, how many creases would there be?

Hint: **Use something to help you.**

Look for a pattern - it helps to **work systematically** and to **keep a record** of what you do.

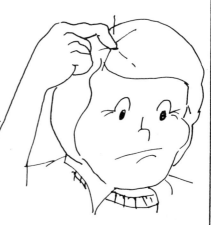

ODDER AND ODDER

What is the sum of the first fifty **odd** numbers?

$1 + 3 + 5 + 7 + 9 + 11 + 13 + 15 + 17 + 19 + 21 + 23 + 25 + ... + 95 + 97 + 99 = ?$

Hint: **Work systematically** and look for a pattern.

Strategies for Problem Solving   Worksheet 7E3-1

REGIONS OF A CIRCLE

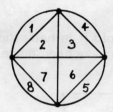

Here is a circle with 4 evenly spaced dots. If we join each dot to every other dot, we divide the circle into 8 regions.

In the table below, I've also recorded the number of regions I found when I had a circle with 2 and with 5 evenly spaced dots.

Number of dots	2	3	4	5	6	7
Number of regions	2		8	16		

Now try it for 3 dots.

Can you see a pattern?

Make a guess for 6 dots. [ ]

Now **count** the regions for 6 dots to **check** your guess.

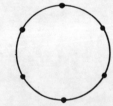

You can try 7 dots too if you like.

Remember: Patterns are useful in problem solving, but by themselves cannot be trusted.

　　　　　　　Patterns need to be **checked.**

　　　　　　　It is also a good idea, after you have checked your pattern, to look back at the problem and try to see **why** the pattern works.

Lesson Block 7F

EXPLAINING WHAT YOU DO

Theme The importance of explaining clearly **what you do.**

Aims To emphasise three important aspects of **explaining what you do** which arise naturally in problem solving:

a) **Explaining Why**

It is important that students realise that finding a pattern or guessing a result is only a step in solving a problem. Before you can trust your result, you need to find a reason <u>why</u> it is true - one which relates the result to the original problem.

b) **Understanding Better**

Explaining what you do requires you to think carefully about what you have done and usually results in significant new insights into the problem. (We all know the saying that "you only really learn a subject when you have to teach it".)

c) **Communication**

Explaining what you do gives practice in communication skills which are as important in mathematics as in other areas, but which are often neglected.

Mathematical Activity

CARTESIAN CHASE and TAKING MATCHES

Instructions for these two games of strategy are given on Worksheets 7F1-1 and 7F1-2.

Summary of Lesson Block

Lesson 7F1 Play CARTESIAN CHASE and TAKING MATCHES and find winning strategies for them (half the class do each game).
"What if..?"s for early finishers.

Lesson 7F2 Explain and check rules and strategies by teaching other class members.

Lesson 7F3 **Explain why** winning strategies work and look at similarities between the two games.

If possible, avoid long gaps between lessons as children forget their strategies.

60   Strategies for Problem Solving

Materials Required and Teacher Preparation

Copies of Worksheets 7F1-1 and 7F1-2 (enough of each for half the class), grid paper, counters (two colours if possible), headless matches.

Play both games with someone else and **find your own winning strategies.**

Plan classroom organisation.
Imagine (or better still, actually have) children seated in two rows of double desks as shown and number children 1, 2, 3 or 4. For Lesson 7F1, children work in pairs 1, 2 and 3, 4. For part of Lesson 7F2, children numbered 2 and 3 swap seats.

```
                    CC              TM
                   ___             ___
                   ___             ___
                   ___   double    ___
                   ___    desks    ___
                   ___             ___
                   ___             ___

              1     2               3     4
```

SAMPLE LESSON PLAN

Lesson 7F1

1. Organise class and hand out Worksheets 7F1-1 and 7F1-2 and materials - half class to play CARTESIAN CHASE, other half TAKING MATCHES.

2. After children have played a few games to learn the rules, introduce the problem: **Find a simple rule to make sure you always win.**

3. When some children think they have a winning strategy, emphasise the need to be able to **explain their strategy clearly to someone else.**

4. Ask children to **write down their winning strategies** as instructions for someone else to follow.

5. As children finish written instructions, ask them to swap instructions (within the same half of the class) and **try to play following the other person's instructions.** **Rewrite** instructions if necessary in view of resulting criticisms.

6. Early finishers can try to find instructions for CARTESIAN CHASE played on boards of other sizes or for TAKING MATCHES with more matches. No child should learn both games yet.

Lesson 7F2

1. Selected children **explain rules of their game** for the other half of the class. Play a few games to learn the rules, at the board or in pairs.

2. At the board, selected children **show that they have a secret strategy** by demonstrating their ability to always win against an opponent from the other half of the class.

   Use pairs 1 v. 3 for CARTESIAN CHASE and pairs 4 v. 2 for TAKING MATCHES. Children **do not** explain their strategy at this stage, only demonstrate it by winning. Make sure selected children know how to win - note likely children at step 4 of Lesson 7F1.

3. Swap children with numbers 2 and 3. Working in pairs, children with number 1 **explain how to always win at CARTESIAN CHASE** to partner with number 3. Similarly, children with number 4 **explain how to win at TAKING MATCHES** to partner with number 2.

It is important that the children doing the explaining have a good strategy - to ensure this, it may be necessary to change the numbers of some children before they swap.

4. At the board selected children **check their learned strategy** by demonstrating their ability to always win against a still uninitiated opponent. (For CARTESIAN CHASE, use pairs of 3 v. 4 and for TAKING MATCHES, pairs of 2 v. 1.)

Lesson 7F3

1. Compare different children's **"rules for winning"** each game. Encourage children to try to **explain why the rules work.**

2. Emphasise the need for clear, concise rules. Ask children to **write down** a "best possible" rule for their own (first) game. (If short of time, include this as homework.)

3. Discuss the similarities in the strategies for the two games. (This may need to be largely exposition by the teacher - see Mathematical Notes.)

4. (Optional) Discuss the results of some of the "What if ...?"s.

5. (Perhaps for homework) Complete write up, including "What problem solving idea I found helpful".

MATHEMATICAL NOTES

CARTESIAN CHASE

Most children find that the squares labelled W are "winning" squares - i.e. a player can always win after putting her/his counter in such a square. Square (3,3) is almost always the first winning square identified.

Most children recognise that the winning squares are characterised by having both "co-ordinates" odd numbers. One child described this by multiplying "co-ordinates" together and noticing that the product must be odd for winning squares.

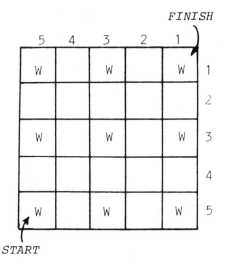

A concise winning strategy for 5 x 5 CARTESIAN CHASE is:
   a) Go first.
   b) Start by placing a counter in the bottom left hand corner.
   c) For all future moves, move in the same direction as your opponent just did.

For a rectangular board of any size, there is a similar winning strategy. If the board is odd x odd choose to go first, thereby starting on a winning square at your first move. Otherwise, let your opponent start and move onto a winning square at your own first move. For all future moves, copy your opponent.

62   *Strategies for Problem Solving*

TAKING MATCHES

This seems more difficult than CARTESIAN CHASE, perhaps because it is harder to keep a record of what was done. If children are **stuck,** suggest they try a smaller game first (3 matches in each pile or even just 2). Many children recognise that a player can always win by leaving an even number of matches in each pile after her/his move.

A concise winning strategy for TAKING MATCHES with four matches in each pile is:

    a)   Go second.
    b)   Always make exactly the same move as your opponent just did.

For any number of matches in each pile, there is a similar strategy. If there are even numbers of matches in <u>both</u> piles, choose to go second. Otherwise, choose to go first and make both piles even with your first move. The number of matches in each pile is kept even by following the strategy above until finally there are no matches in either pile.

"WHAT IF...?"S

Before attempting more difficult "What if...?"s, children should find strategies for different size games on square boards (or with equal piles of matches). Rectangular boards and unequal piles provide further natural extensions.

Rules regarding allowable moves could also be altered - for example, more than one match could be removed from each pile. However, the method of solution will no longer carry over in such an obvious way.

BOTH GAMES ARE REALLY THE SAME!

You begin to suspect this when you note the similarities in the winning strategies. But it is well disguised by the different formats for the games and also because 5 x 5 CARTESIAN CHASE is the same as 4,4 TAKING MATCHES. To illustrate why the games are the same, you can draw all possible moves for 4,4 TAKING MATCHES in an array, similar to 5 x 5 CARTESIAN CHASE.

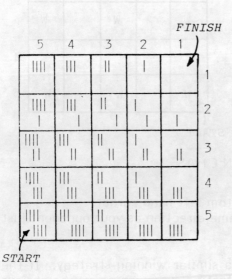

Notice that the allowable moves for TAKING MATCHES are in the same directions as for CARTESIAN CHASE and that the winning positions (of leaving both piles even) correspond exactly to the winning squares in CARTESIAN CHASE.

Strategies for Problem Solving · Worksheet 7F1-1

CARTESIAN CHASE

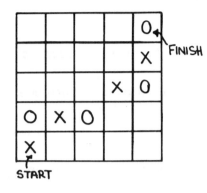

CARTESIAN CHASE can be played by 2 players on a 5 x 5 board as follows: Take it in turns to place counters on the board, starting at bottom left hand square.

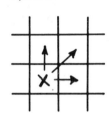

If your opponent places a counter in the square marked X here, you must place your counter in one of the 3 squares indicated by the arrows.

Player X starts, Player 0 wins.

The winner is the player who places her counter in the top right hand square.

**YOU CAN CHOOSE TO PLAY FIRST OR SECOND.
CAN YOU FIND A WAY OF ALWAYS WINNING?**

Strategies for Problem Solving · Worksheet 7F1-2

TAKING MATCHES

Start with two piles of four matches or counters. The two players take it in turns to remove matches. Each player may remove **either** one match (from either pile) **or** one match from both piles.

The winner is the player who removes the last match.

Here is an example

Player 1	↑ ↑ ↑ ✗	✗ ↑ ↑ ↑
Player 2	↑ ↑ ✗	✗ ↑ ↑
Player 1	↑ ✗	↑ ↑
Player 2	✗	↑ ↑
Player 1		✗ ↑
Player 2		✗

PLAYER 2 WINS

**YOU CAN CHOOSE TO PLAY FIRST OR SECOND.
CAN YOU FIND A WAY OF ALWAYS WINNING?**

Lesson Block 7G

SPIROLATERALS

Theme	**Examining special cases.**
Aims	To show that mathematical results are discovered through **a spiral of guessing and checking.**
	To show how important **examining special cases** is to both guessing and checking. It inspires guessing and is central to checking.
	To give children the opportunity to follow their own ideas and create their own hypotheses.
	To practise **'explaining what you have done so someone else can understand'.**

Mathematical Activity

SPIROLATERALS

A spirolateral is a geometric design made from a sequence of numbers. The problem is to investigate some of their many fascinating properties.

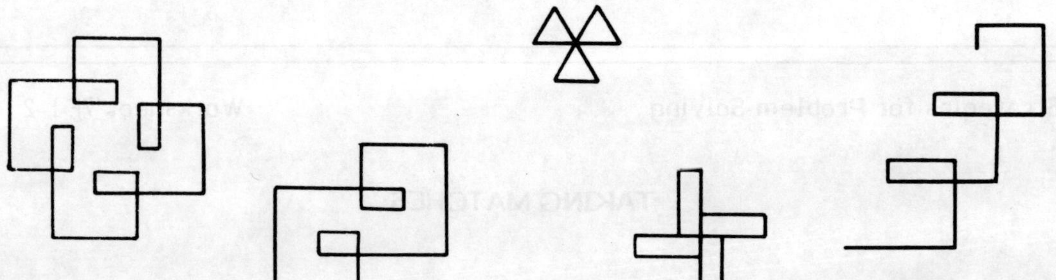

Summary of Lesson Block

Lesson 7G1   Learning to draw spirolaterals and then guessing some of their properties.

Lesson 7G2   Checking and refining the guesses (may be more than one period).

Lesson 7G3   Preparing a report on the discoveries.

Lesson Block 7G 65

Materials Required and Teacher Preparation

Teachers can read about spirolaterals from one or more of the references given in the Mathematical Notes below, but should definitely carry out a personal **investigation of spirolaterals**, with or without the references. Some interesting areas for investigation for both teachers and students are given below.

Students need plenty of **grid paper** (approximately 0.5 cm grid is suitable) for all lessons. **Isometric grid paper** is required for fast workers - **Worksheet 7G2-1** is a master. **Project materials** (pens, cardboard etc) are required for preparing a report in Lesson 7G3.

A **grid chalkboard** with contrasting chalk or an **overhead projector** with a grid are useful for displaying spirolaterals to the class.

Exciting possibilities exist if children have access to a **micro-computer and the LOGO turtle.**

SAMPLE LESSON PLAN

Lesson 7G1

1. Decide firmly if spirolaterals are going to turn **left or right.**

2. Ensure all children have an adequate supply of grid (squared) paper. Introduce spirolaterals as a way of making attractive designs from number sequences. (One teacher brought in a children's picture book about a worm and talked about imagining the patterns a worm would leave behind in the soil.)

3. HOW TO DRAW A SPIROLATERAL

   Take any sequence of integers and, if it is of finite length, imagine it repeated indefinitely. Imagine a worm (or LOGO turtle) at the centre of the grid, facing 'north'. If the sequence is (p, q, r, s, ...) the worm moves p units forward, turns right by $90°$, moves q units forward, turns right by $90°$, moves r units forward, turns right by $90°$, moves s units forward, turns right by $90°$, etc.

   Here are some examples. The infinite sequence (1, 2, 3, 4, 5, ...) spirals outwards. The finite sequence (1, 2, 3, 4) when repeated indefinitely produces an infinite string of loops. The finite sequence (1, 2, 3) closes back on itself after four repetitions, producing a 'windmill'.

   (1, 2, 3, 4, 5 ...)   (1, 2, 3, 4)   (1, 2, 3)

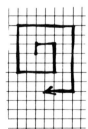 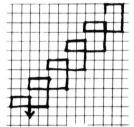 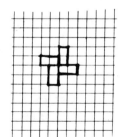

4. Give children some spirolaterals to draw, for example (3, 4), (4, 1, 2), (1, 2, 1, 3, 1, 4, 1, 5, 1, ...), (1, 3, 2, 6), (1, 2, 1, 2, 1).

   Insist all spirolaterals drawn now and later are labelled and preserved for future reference.

   Many students will make mistakes in drawing spirolaterals because they do not appreciate that the 'worm' always turns right with respect to itself; instead there is a tendency to turn right with respect to the paper producing results like this:

   (1, 2, 3)

   To overcome this, some students may find it helpful to turn the paper so the worm is always heading 'north'. Try to ingrain the physical feel of always turning right by first practising drawing squiggles like these:

   Watch out for these difficulties and try to eliminate them now.

5. Ask children to **draw personalised spirolaterals**, say their home postcode, telephone number or date of birth (18 July 1971 becoming (1, 8, 7, 1, 9, 7, 1) for example). Again insist on labelling for future reference. Students may like to use these personal spirolaterals for greeting cards, book covers etc.

   The question of what to do with zero will arise. It can be ignored, but it is better to interpret it as a 90° turn with no movement. As shown below, the two different interpretations result in different patterns.

   (1, 2, 0, 3)   (1, 2, 0, 3)

   (with zero ignored)   (making movement of zero length)

6. While some children put their spirolaterals on the board or overhead projector, ask the rest of the class to guess the generating number sequences. Note that more than one sequence is often possible.

   Encourage class discussion of these examples to **bring out properties of spirolaterals** - for example, some designs are (geometrically) similar, some designs (such as most from seven-digit telephone numbers) 'close' whereas others (such as most from four-digit postcodes) are infinite. Try to get these observations and others **coming from the class,** not the teacher.

7. In the last 10 minutes, explain that children are to **investigate any aspect of spirolaterals that they find intriguing** and are to present their discoveries as a project. They may work alone or in small groups. Before the end of the class, each child should **jot down some guesses about the behaviour of spirolaterals,** so productive checking and refining of guesses can begin next time. **See the Mathematical Notes for some profitable avenues of investigation** - be careful not to push children to ask questions that are too general and encourage them to make their own observations and investigations rather than relying on the teacher's suggestions only.

Lesson 7G2

1. Begin by reminding children of what a spirolateral is and warning them to 'keep right'. Get a few children to read out the guesses written down last time. Explain, and summarise prominently on the board, **how the spiral of guessing and checking works in discovering mathematical properties.**

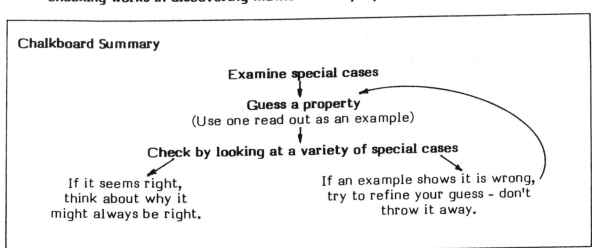

2. Each child should **write down the exact question** he or she is investigating before work starts - this is to clarify the idea in the student's mind. As they begin work, the teacher should read this and **make suggestions for broadening, restricting or clarifying,** as appropriate.

3. For those students progressing quickly, suggest the possibility of generalising their results to **spirolaterals which turn through angles other than 90°.** Isometric paper is convenient, allowing turns of either 120° or 60°. Worksheet 7G2-1 is a master.

(1, 2, 3) turning through 120°

(1, 2, 3) turning through 60°

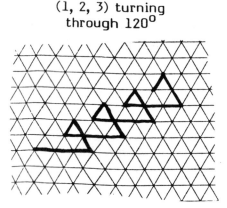
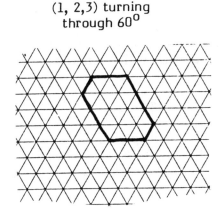

Lesson 7G3

1. In this lesson, children should write down what they have discovered **'so that someone else can understand'**. This could be in the form of a wall-chart project to be begun during this lesson.

   Stress the need for:
   - Explaining clearly what a spirolateral is. (Perhaps one group could be allocated this task).
   - Explaining exactly what questions the project answers.
   - Presenting results clearly - probably with examples.

2. Complete projects for homework and display for other people to see.

MATHEMATICAL NOTES

Teachers may wish to consult one or more the the following references.

Olds, Frank C., **'Spirolaterals'**, Mathematics Teacher, 66 (Feb. 1973), pp 121-124.
Schwandt, A.K., **'Spirolaterals: Advanced Investigations from an Elementary Standpoint'**, Mathematics Teacher, 72 (March 1979), pp 166-169.
'Worm Patterns', Australian Mathematics Teacher, Vol 36, No 2 (1980), pp 16-20.
Mottershead, L.J., **'Spirolaterals'**, L and S, Melbourne, 1979.

SOME POSSIBLE AVENUES OF INVESTIGATION

Teachers should encourage children to begin by investigating fairly restricted questions and, if possible, generalise them later. The many labelled spirolaterals created in the first lesson should form a bank of special cases for guessing and checking. The questions below are not prescriptive, but are provided to indicate some rich areas for investigation.

Question 1   What is the pictorial relationship between spirolaterals generated by related number sequences?

For example, what happens to the spirolateral if I double each number, triple each number, add one to each number etc.? What is the relationship between the spirolaterals belonging to (1), (1, 2, 3), (1, 2, 3, 4), (1, 2, 3, 4, 5), ...?

Question 2   What other number sequences make the same spirolateral as my telephone number does?

Question 3   All spirolaterals from sequences of length 2 are rectangles. Do all sequences of length 3 have 'windmill' spirolaterals? What designs can be made by sequences of other specified lengths.

SOME SKETCHY ANSWERS

Question 1   Multiplying all the terms in a number sequence by a constant results in geometrically similar spirolaterals. Adding a constant to all terms does not change the basic shape, although they are no longer geometrically similar.

Question 2   Many sequences can generate one spirolateral. For example the windmill generated from (1, 2, 3) can be generated by (2, 3, 1) and (3, 1, 2) although the starting points are different. However there are other possibiities such as (2, 1, 2, 3, 1, 2, 3, 1, 2, 3, 1, 2, 1).

Question 3   All sequences consisting of three numbers have 'windmill' spirolaterals although there are several types, as well as rotations and reflections.

(1, 1, 1)   (1, 2, 3)   (1, 2, 5)   (3, 3, 1)
(A degenerate
 windmill!)

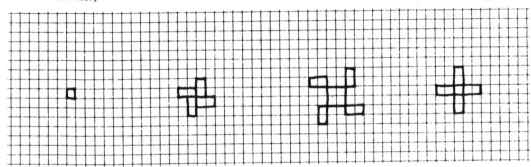

After four repetitions, the spirolateral made from (p, q, r) is back to its starting point and heading 'north' again, so that the four 'blades' of the windmill will be traced over and over again.

Movements made in first four cycles
p north, q east, r south, p west
q north, r east, p south, q west
r north, p east, q south, r west

∴   Total movement north = p + q + r = total movement south
       Total movement east = p + q + r = total movement west
This means that the spirolateral is back to its starting point.

A similar analysis applies to all sequences with an odd number of terms - after four repetitions they must return to the starting point and be heading north ready to trace the pattern over again.

Examples:

Movements made in first four cycles
     for (p, q, r, s, t)

North	East	South	West
p	q	r	s
t	p	q	r
s	t	p	q
r	s	t	p
q	r	s	t

Movements made in first four cycles
   for (p, q, r, s, t, u, v)

North	East	South	West
p	q	r	s
t	u	v	p
q	r	s	t
u	v	p	q
r	s	t	u
v	p	q	r
s	t	u	v

In each case the total movement north is now equal to the movement south and the total east is equal to the total west.

70 *Strategies for Problem Solving*

For sequences with an even number of terms, the situation is more complicated, as the moves of each length do not occur in each direction. For example, the moves made by the sequence (p, q, r, s) are always p units north, q east, r south and s west. The spirolateral will therefore only close if $p = r$ and $q = s$, in which case it forms a rectangle. Otherwise it will continue indefinitely. As a second example, consider the 6 term sequence (p, q, r, s, t, u). It closes after only two repetitions, when the total movements north and south are both $p + r + t$ and the total movements east and west are both $q + s + u$. This is typical of even numbers that are not multiples of four. When the number of terms is a multiple of four, the spirolateral will either continue indefinitely or close if there is a special relationship between its terms. For example the 8 term sequence (p, q, r, s, t, u, v, w) moves $p + t$ north, $q + u$ east, $r + v$ south and $s + w$ west each cycle. Hence it will close if and only if $p + t = r + v$ and $q + u = s + w$. Two types of spirolateral with eight terms are shown below.

(1, 2, 3, 4, 5, 6, 3, 4)   (1, 2, 3, 4, 5, 6, 7, 8)

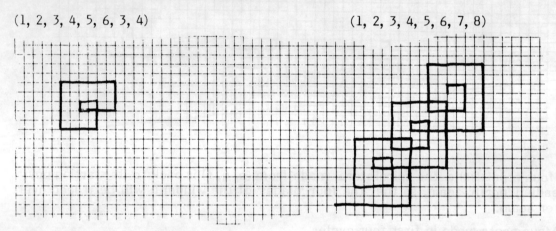

Below are listings for two MIT LOGO procedures which, together, allow spirolaterals to be drawn on the Apple computer. To run the program, type SPIROLATERAL. When asked, the user must choose the angle at which the TURTLE turns and, later, enter the number sequence with a space between each number. To allow for the possibility of spirolaterals closing or continuing indefinitely, neither procedure contains a command to STOP. (This is obvious when drawing spirolaterals which continue indefinitely, but gives the appearance of the computer having stopped, with no prompt appearing, when drawing spirolaterals which close - in fact, the turtle is retracing its path indefinitely.) To stop the program, use CTRL-G. To draw another spirolateral, retype SPIROLATERAL.

```
TO SPIROLATERAL
  PRINT [PLEASE ENTER THE ANGLE YOU WANT TO USE]
  MAKE "ANG FIRST REQUEST
  PRINT [PLEASE ENTER YOUR SEQUENCE OF NUMBERS]
  PRINT [LIKE THIS 2 3 9 6]
  MAKE "N REQUEST
  FULLSCREEN
  DRAW
  HIDETURTLE
  WORM :N
END
```

```
TO WORM :M
  TEST :M = []
  IFTRUE MAKE "M :N
  MAKE "L FIRST :M
  FD :L * 5 RT :ANG
  WORM BUTFIRST :M
END
```

Strategies for Problem Solving   Worksheet 7G2-1

ISOMETRIC SPIROLATERALS

On this paper spirolaterals which turn through 60° or 120° at each step can be easily drawn.

Lesson Block 7H

GENERALISING

Theme
: General rules are discovered by articulating the patterns observed in many particular cases.

Aims
: To give experience of several successive layers of generalising;

> from one very particular case,
> to a rule,
> to a much more widely applicable (more general) rule.

To consolidate many aspects of problem solving involved in earlier lessons, in particular

> looking for patterns,
> using trial and error to attack problems,
> working systematically,
> recording sensibly,
> explaining what you have done.

Note: While these lessons complement the teaching of algebra, DO **NOT** USE ALGEBRA TO FIND THE SOLUTIONS. However, with a suitable class, a teacher may want to encourage the natural use of algebra at the end when stating the "rule".

Mathematical Activity

STAY-THE-SAME NUMBERS

The number 5 is a "stay-the-same" number because it doesn't change when it is put into this function machine, since 5 × 4 = 20 and 20 - 15 = 5.

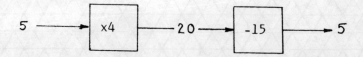

Find a rule to predict the stay-the-same numbers for other function machines which first multiply and then subtract.

Summary of Lesson Block

Lesson 7H1 After an introduction or revision of function machines, children do examples of STAY-THE-SAME NUMBERS for function machines of the type

Children try to find a rule for this type, using guessing and checking.

Lesson 7H2 Class works in small groups to find rules for other function machines. Through a teacher-led discussion, the whole class tries to **find a general rule.**

Lesson 7H3 Stay-the-same numbers can be found for a wider class of function machines.

Materials Required and Teacher Preparation

Teachers should first try the problem **without using algebra,** taking care to note which types of examples result in stay-the-same numbers that are natural numbers, fractions or negatives. The worksheets provided (Worksheets 7H1-1, 7H2-1, 7H2-2, 7H2-3) include examples with answers of all these types, so some teachers may wish to replace some examples by easier ones. Each child requires a copy of Worksheet 7H1-1, but only one of the three Worksheets 7H2-1, 7H2-2 or 7H2-3.

For Lesson 7H3, calculators may be useful.

SAMPLE LESSON PLAN

Lesson 7H1

1. **Introduce (or revise) function machines,** giving a variety of examples and showing what they do to various numbers.

    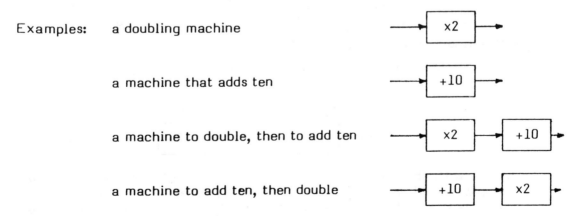

    In this lesson, all the function machines will first multiply then subtract.

74 Strategies for Problem Solving

2. Hand out Worksheet 7H1-1 and give some examples of stay-the-same numbers. For example

   5 is a stay-the-same number for →[x4]→[-15]→

   7 is a stay-the-same number for →[x3]→[-14]→

3. Ask children to **work individually or in pairs finding stay-the-same numbers** for function machines of the form →[x3]→[-n]→, as on Worksheet 7H1-1.

   Emphasise possible strategies such as **guess and check, trial and error, working systematically and looking for patterns.**

4. Ask children to find, and write down as clearly as possible, **a rule which will predict the stay-the-same numbers** for function machines which first multiply by 3. Make sure children test their rules by making predictions and checking them.

Lesson 7H2

1. Discuss rules written down in step 4 above. Most children have probably written down something like "the stay-the-same number is half the number you subtracted". **Examine the validity of this** (or other suggested rules) by examining special cases.

2. Challenge the class to generalise this rule so that it works for function machines that multiply by numbers other than three.

   To start, **divide the class into six groups**, to examine x2, x4, x5, x7, x10, x11 examples. Ask each group to **write down a rule** for their type of function machine. See Worksheets 7H2-1, 7H2-2 and 7H2-3.

3. Through teacher-led discussion, **collate the rules** found by each of the six groups to create a general rule for any function machine →[xm]→[-n]→.

   Children might express it like this: to find the stay-the same number you take the number subtracted by the function machine and divide it by one less than the multiplying number. This works unless the multiplying number is one.

   With suitable classes, algebra can be used to express this result more succinctly.

   $$\text{Stay-the-same number} = \frac{n}{m-1}$$

4. Check the rule by using it to predict answers to new examples, such as

   →[x31]→[-120]→

   Discuss with the class whether or not the rule definitely works - **how many examples would we have to try to be really sure?** Some children (who cannot prove the result algebraically) will nevertheless be able to give a general reason why the result is true. A few children may be able to formulate the problem and prove the result algebraically but **do not push** this when it does not yet come naturally.

Lesson 7H3

1. **"What if ...?"s.**

    Ask for suggestions from the class and select some promising ones for students to try. For example,

    (i)   What if we reversed the order so that we subtract and then multiply?
    (ii) What if we add instead of subtracting?
    (iii) What if we divide instead of multiplying?
    (iv) What if we try squaring first, then subtracting like these:

    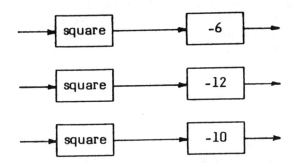

2. Make sure children write down their results "so that someone else can understand".

3. Finish the lesson with brief reports on the results obtained.

MATHEMATICAL NOTES

CHILDREN'S SOLUTIONS

While this may appear to be an unappealing problem to people familiar with algebra, it is a very nice problem for children who tend to **look for patterns.**

For example,

results in some children setting up a table similar to

Number	x3	-16
10	30	14
11	33	17
12	36	20
.	.	.
.	.	.
.	.	.
8	24	8

76 Strategies for Problem Solving

Many patterns can be observed such as "Whenever the number increases by 1, the final number increases by 3". These allow a more systematic approach to **trial and error**.

Many children observe partial "rules" such as "the stay-the-same number must be bigger than 5, because we need to be able to take away 16".

Similar patterns and rules emerge when children try to find the general rule by combining their "data" for the various cases of x3, x 4, etc.

EXAMPLES INVOLVING SQUARES

Examples involving squares give rise to interesting patterns Whereas the function machine that squares and then subtracts 6 has two stay-the-same numbers (3 and -2), the machine that squares and then subtracts 10 has no integer or fractional stay-the-same number. Its stay-the-same numbers can be shown algebraically to be

$$\frac{1 + \sqrt{41}}{2} \text{ and } \frac{1 - \sqrt{41}}{2}$$

Children could find numerical approximations to these. Examples such as these can lead to many interesting discussions, especially when calculators are readily available.

Strategies for Problem Solving   Worksheet 7H1-1

STAY-THE-SAME NUMBERS

Look what happens when 6 is put into this function machine

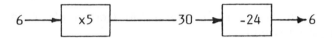

We call 6 a STAY-THE-SAME number.

Find the stay-the-same number for this function machine

As you work, record your guesses and their results here.

Guess (IN)	10	20	1						
Result (OUT)	16	46	-11						

Now write the stay-the-same number in the triangle in the table below.

Number being subtracted	14	16	18	20	1	2	3	4	5	6	100
Stay-the-same number	△										

Find the stay-the-same numbers for the function machines below and record them in the table above.

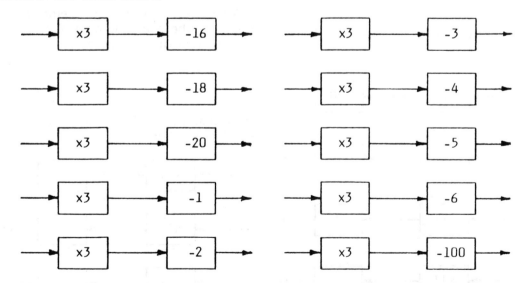

Write down a rule for predicting the stay-the-same number here.

Strategies for Problem Solving   Worksheet 7H2-1

STAY-THE-SAME NUMBERS

Find the stay-the-same numbers for these function machines and write them in as shown on the first example.

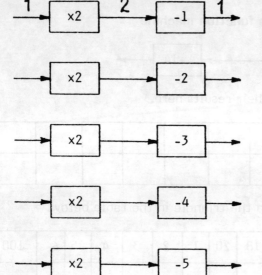

Write down a rule for predicting the stay-the-same number for this type of function machine.

Now test your rule, using your own machines below.

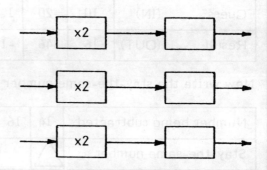

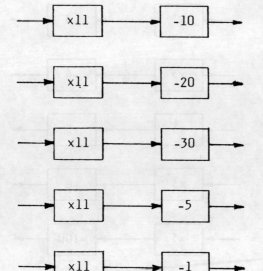

Write down a rule for predicting the stay-the-same number for this type of function machine.

Now test your rule, using your own machines below.

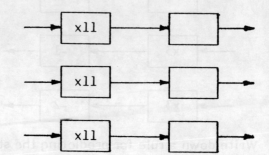

Strategies for Problem Solving — Worksheet 7H2-2

STAY-THE-SAME NUMBERS

Find the stay-the-same numbers for these function machines and write them in as shown on the first example.

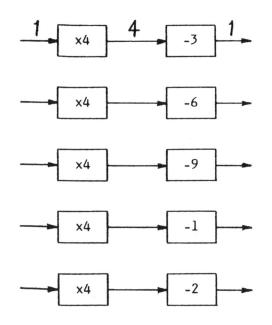

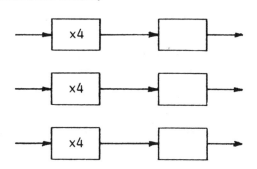

Write down a rule for predicting the stay-the-same number for this type of function machine.

Now test your rule, using your own machines below.

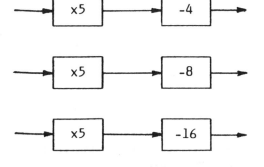

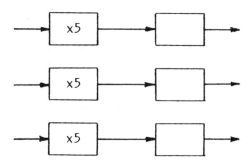

Write down a rule for predicting the stay-the-same number for this type of function machine.

Now test your rule, using your own machines below.

Strategies for Problem Solving   Worksheet 7H2-3

STAY-THE-SAME NUMBERS

Find the stay-the-same numbers for these function machines and write them in as shown on the first example.

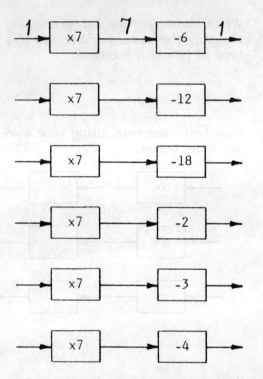

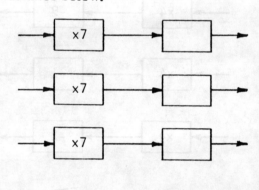

Write down a rule for predicting the stay-the-same number for this type of function machine.

Now test your rule, using your own machines below.

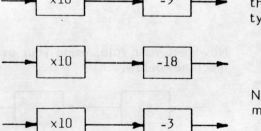

Write down a rule for predicting the stay-the-same number for this type of function machine.

Now test your rule, using your own machines below.

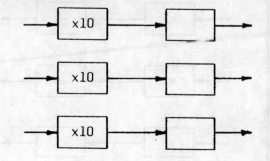

Lesson Block 8A

LEAPFROGS

Theme   Use something to help you.

Aims   To show a variety of tools that can be helpful in problem solving, including

> **concrete aids** (e.g. counters),
> **diagrams** (e.g. showing how the counters move),
> **tables and charts,**
> **mathematical tools for looking for patterns and expressing answers.**

To provide a **general introduction to the second year of the problem solving course** by showing children how written work is to be presented and by demonstrating several problem solving strategies, such as

> using concrete aids and diagrams
> working systematically
> developing a good recording system
> collecting and organising data
> recognising and extending patterns.

Mathematical Activity

LEAPFROGS

On a row of 5 squares, 2 red counters and 2 black counters are placed like this.

The counters can slide one place (either right or left) into a vacant space or they can hop over a counter, again into a vacant space.

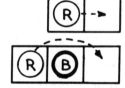

Can you finish with the black counters where the red ones were and the red counters where the black ones were?

What would be the least number of moves needed?

82 Strategies for Problem Solving

Summary of Lesson Block

Lesson 8A1   Do LEAPFROGS puzzle with 2 and 3 counters of each colour.

Lesson 8A2   Solve a more general LEAPFROGS puzzle by collectively gathering evidence and finding a pattern. Again the teacher should help children decide what to report and how to present written work.

Materials Required and Teacher Preparation

Children will need counters (or substitutes) of two colours for frogs, but they can easily draw their own boards. To demonstrate moves to the class, it is helpful to have magnetic counters on the chalkboard or counters of different shapes to move on the overhead projector. The teacher should become familiar with the patterns involved in the puzzle, both in the moves of the counters and the resulting number patterns.

SAMPLE LESSON PLAN

Lesson 8A1

1. Introduce the LEAPFROG puzzle with 2 counters of each colour.

   Suggest that children draw up their own boards (make sure they are large enough for the counters) and then play individually or with a neighbour. Whilst one child moves the counters, the partner can check, count and record.

   It usually happens that **children soon find out how to make the interchange but then forget how they have done it.** Allow sufficient time here, during which the teacher can circulate, check strategies and encourage children to keep a record of their moves.

   Suggest to individuals who finish quickly that they could try LEAPFROGS with 3 counters and then 4 counters of each colour.

2. When most children have been able to interchange the counters (and have written down how to do it) **take suggestions for least number of moves** (answer: 8) and ask some children to **demonstrate how it is done.** The magnetic counters or overhead projector are useful here.

   Three points can emerge during these demonstrations.

   (a) The configurations $\boxed{R|R|B}$ and $\boxed{R|B|B}$ cause **blockages** that must be "backed out of", thereby wasting moves.

   (b) In the method that uses the least number of moves, **none of the counters has to move backwards** (i.e. in the above examples, reds always move right and blacks always move left.)

   (c) There are two methods of performing the interchange in the least number of moves, but **they are essentially the same;** one moves a red first, the other a black.

3. The children explaining their methods on the board will have used a variety of notations - ways of recording their moves. **Discuss the importance of choosing a convenient system and discuss advantages and disadvantages of the systems suggested.** Allow children to choose personal favourites for later use - below are three common systems.

Lesson Block 8A   83

Example 1

(WITH NO "BACKWARDS" MOVES)

Move	R	R		B	B	START
1	R	R	B		B	(either red or black can move first)
2	R		B	R	B	(only possible move as a black slide causes blockage)
3		R	B	R	B	(alternative causes blockage after next move)
4	B	R		R	B	(only possibility)
5	B	R	B	R		(alternative of red slide would cause blockage)
6	B	R	B		R	(only possibility)
7	B		B	R	R	(only possibility)
8	B	B		R	R	(we're home!)

Advantage: a complete record of moves taken.
Disadvantage: too long if you don't want completeness.

Example 2

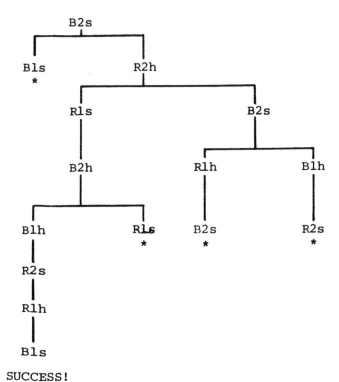

Notation: The counters are labelled R1, R2, B1 and B2 and the starting position is

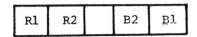

A slide is denoted by s and a hop by h so B2s means a slide by counter B2 and R1h means a hop by counter R1 etc.

No backwards moves are being considered; * denotes a position where progress cannot be made without a backwards move.

SUCCESS!

Advantages: a complete record of all moves considered.
Disadvantage: hard to keep track of the position of the counters.

Example 3

Move Number	Move Type
1	B slide
2	R hop
3	R slide
4	B hop
5	B hop
6	R slide
7	R hop
8	B slide

Total: 8 moves consisting of 4 slides and 4 hops.

Note the symmetry of moves (1 with 8, 2 with 7 etc).

Advantages: Simple and compact. Surprisingly easy to follow with counters.

Disadvantages: Only provides hop/slide information. Doesn't keep track of individual counters.

Example 3 more briefly: Bs, Rh, Rs, Bh, Bh, Rs, Rh, Bs.

4. Tell children of four features expected of written work as explained in the Practical Teaching Tips. These headings could be written inside the front cover of the children's books for easy reference.

   (i) **Name of problem and clear statement of aim.**

   (ii) **Rough working:** NO RULES ABOUT THIS

   (iii) **What I discovered.** (A brief summary of the results)

   (iv) **What ideas I found helpful.** (One or two things that really made a difference.)

   Compose (with the class) something to write under the latter two headings. In later lessons, children should become more sensitive to what helps them solve problems and will need less help. Perhaps they might come up with something like this:

   What I discovered

   I found that 8 moves is the least number. I do it this way (display method using preferred notation). I never move backwards; that's why this gives the least number of moves.

   What ideas I found helpful

   The counters. My final system for writing things down. Noticing how two reds or two blacks together caused a blockage.

5. Allow time for writing this summary or set it for homework. Also set children the LEAPFROGS problem with 3 counters on each side. Suggest that **searching for a pattern in the sequence of moves in the 2 counters case** might help with 3 counters. Suggest that they use one of the recording systems developed earlier.

Lesson 8A2

1. Recall LEAPFROGS problem of Lesson 8A1 and the answer for 2 counters on each side and for 3.

   Introduce the following challenge: **What if the number of leapfrogs on each side is changed - how can we predict how many moves will be needed to**

perform the interchange? For example, how many moves would be needed if there were 10 blacks and 10 reds or perhaps 17 reds and 83 blacks?

2. **Formulate a class plan** for answering the challenge. Ten counters on each side is too big to tackle directly so we will **work systematically** on small cases, hoping to find a pattern from which we can **confidently predict** the answer we need.

3. Divide the work amongst class members, first setting one quarter of the class to work on 1 red and 1 black, then 1 red and 2 blacks, 1 red and 3 blacks etc.

    Set the other quarters to work on

    - 2 reds and various numbers of blacks (1,2,3,...)
    - 3 reds and various numbers of blacks
    - 4 reds and various numbers of blacks.

    Offer this hint: **the pattern will be easier to find if you record separately the number of hops and the number of slides.**

    Stress that children should **work systematically on small cases.**
    Stress that they should **look for patterns and check them.**

4. While they are working, put up on the board all results previously found by the class. (There aren't many.) It is easier to spot the pattern if separate tables are constructed for hops and slides as shown in Figures 1a and 1b.

NUMBER OF SLIDES

REDS	BLACKS			
	1	2	3	4
1	2	3	4	5
2		4		
3			6	
4				

Figure 1a

NUMBER OF HOPS

REDS	BLACKS			
	1	2	3	4
1	1	2	3	4
2		4		
3			9	
4				

Figure 1b

5. Collate results from the first quarter of the class to construct the tables illustrated in Figures 1a and 1b. Discuss the patterns evident here. See if the patterns hold as the results of other quarters are added, gradually building up to Figures 2a and 2b. Try to modify them so they do work, rather than looking for entirely new patterns.

NUMBER OF SLIDES

REDS	BLACKS			
	1	2	3	4
1	2	3	4	5
2	3	4	5	6
3	4	5	6	7
4	5	6	7	8

Figure 2a

NUMBER OF HOPS

REDS	BLACKS			
	1	2	3	4
1	1	2	3	4
2	2	4	6	8
3	3	6	9	12
4	4	8	12	16

Figure 2b

6. Look at results from all groups to find general pattern (that with m reds and n blacks there are m + n slides and m x n hops). Some more able students will be able to explain why these patterns hold, especially the numbers of hops. Use the pattern to answer the question about 10 counters on each side and other, larger numbers.

7. Finish the lesson with a summary of all the things that were used to make the investigation easier: working systematically, the ways of recording moves, the tables and anything else.

8. Get the children to write what they have found out about LEAPFROGS. Allow sufficient time for them to do this carefully and clearly or set it for homework.

9 Extension problems: High fliers could be asked to explain the result for the number of hops and investigate how the number of moves changes when there is more than one vacant space.

MATHEMATICAL NOTES

THREE FROGS ON EACH SIDE

Move	R	R	R		B	B	B
1	R	R	R	B		B	B
2	R	R		B	R	B	B
3	R		R	B	R	B	B
4	R	B	R		R	B	B
5	R	B	R	B	R		B
6	R	B	R	B	R	B	
7	R	B	R	B		B	R
8	R	B		B	R	B	R
9		B	R	B	R	B	R
10	B		R	B	R	B	R
11	B	B	R		R	B	R
12	B	B	R	B	R		R
13	B	B	R	B		R	R
14	B	B		B	R	R	R
15	B	B	B		R	R	R

Moves 1–4: B slide, R hop, R slide, B hop — First 4 moves of (2,2) frogs puzzle.

Moves 5–11: B hop, B slide, R hop, R hop, R hop, B slide, B hop — Symmetric 5 – 11, 6 – 10, etc.

Moves 12–15: B hop, R slide, R hop, B slide — Last 4 moves of (2,2) frogs puzzle.

Summary:
15 moves
9 hops
6 slides

Note the symmetry of the moves again (1 with 15, 2 with 14, etc.) Again, there are only 2 ways of doing the puzzle in the least number of moves, one moves a red first, the other a black.

THE GENERAL CASE - FOR TEACHERS

The number of hops is easy to find. If there are m red frogs and n black frogs, then the m red frogs must each hop over or be hopped over by each of the n black frogs. This means there are $m \times n$ hops needed. (So when $m = n = 2$, there are 4 hops and when $m = n = 3$ there are 9 hops.) Some children will be able to appreciate this argument.

To find the number of slides, first consider the total number of spaces to be moved. The m red frogs each have to move $n+1$ spaces and the n black frogs each have to move $m+1$ spaces so there are $m(n+1) + n(m+1)$ spaces to be

88 Strategies for Problem Solving

moved altogether. But this is equal to the number of slides plus twice the number of hops, as each hop accounts for 2 spaces moved.

Number of spaces to move	= number of slides + 2×(number of hops)
m(n+1) + n(m+1)	= number of slides + 2×mn
number of slides	= m+n
Also number of moves	= number of slides + number of hops
	= mn + m + n.

Children can guess the results on numbers of hops and slides separately by looking at the number patterns in suitable tables and then use the separate results to find the total number of moves.

More than one space in the middle makes a difference only to the number of slides. If there are v vacant spaces in the middle, then the number of spaces to be moved is m(n+v) + n(m+v), producing (by the argument above) v(m+n) slides and mn + v(m+n) moves altogether.

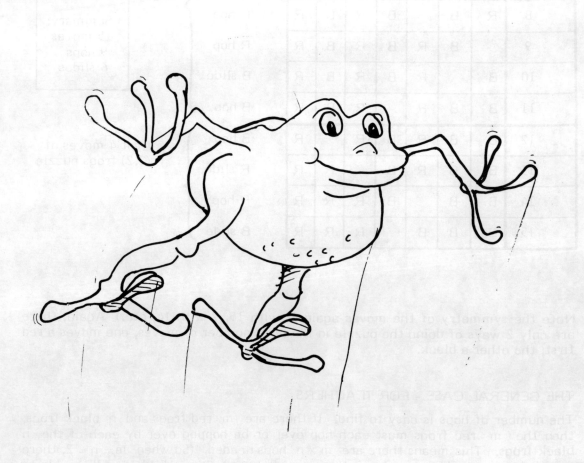

Lesson Block 8B

TRIAL AND ERROR

Theme   **Making trial and error a powerful technique.**

Aims   To show that trying something – almost anything – is a reasonable way to get started on a question.

To teach simple techniques (of being systematic and recording) which can transform into a powerful technique the "hit and miss" trial and error that children spontaneously use.

Children naturally use trial and error as a problem solving method, but usually haphazardly and inefficiently. To make it effective it is necessary

(i)   to do it systematically,

(ii)   to keep a written record of the trials and the results (the errors),

(iii)   to examine the errors to choose the next trial.

This basic idea is very important as it leads on to iterative methods such as Newton's method of solving equations. Although these methods have always been important, calculators and computers now make them particularly effective.

Mathematical Activity

GUESSING GAMES

The trial and error strategy is introduced by a simple guessing game. A variety of short number and geometric problems are used to reinforce the ideas.

Summary of Lesson Block

Lesson 8B1   Introduce, using the guessing game, the trial and error strategy outlined above and use it to solve some number puzzles with the whole class. Suggest helpful ways of recording the trials and errors.

Lesson 8B2   Practise the trial and error strategy on problems selected from Worksheet 8B2-1.

Materials Required and Teacher Preparation

The teacher should first try the problems using a trial and error strategy rather than algebra. Calculators are useful for both lessons. By speeding up the calculations, calculators make the trial and error process much more palatable, but they are not essential. Teachers may wish to distribute individual copies of Worksheet 8B2-1 or use the overhead projector.

SAMPLE LESSON PLAN

Lesson 8B1

1. Explain rules for the guessing game as follows. Someone (initially the teacher) thinks of a positive whole number less than 1000. Others guess what it is and are told if the number is greater than, less than, or equal to the guess. Record the number of tries needed to guess the number.

    The best method is called a **binary search** and is commonly employed by computers. Make the first guess the middle number in the interval - i.e. start with 500 then go to 250 or 750 depending on the reply. Continue in this fashion until the number is located. This technique ensures that only 10 guesses are necessary because $2^{10} > 1000$. The average number of guesses is much less than 10.

    Some students will use this method - it can be explained to the class if desired - but students should not be given the impression that "halving" is always best in a trial and error situation.

2. Play the game two or three times, making it harder (if appropriate) by allowing decimals. **For example, try to find a decimal number between 0 and 1.** Try to make the game hard enough that keeping a record of the trials (i.e. guesses) is beneficial.

    Point out the essentials of the strategy and write them prominently on the chalkboard. Explain how this strategy is being used in the guessing game and that it is also useful in many other types of problems.

3. Show how the guessing game strategy can be used to solve other types of problems, for example NUMBER PUZZLE A. **Demonstrate the use of a well-labelled table** of trials and errors, such as is given in the Mathematical Notes.

4. Spend any remaining time in this period on some of the other Number Puzzles. They are arranged in order of difficulty. Allow children to work with calculators at their own pace, but display at least one child's table of trials and errors for each problem and discuss **exactly how the errors were used to choose the next trial.**

Lesson 8B2

1. Begin this lesson by revising the trial and error strategy and again putting up the chalkboard summary suggested above.

    Allow students to work on selected problems from Worksheet 8B2-1 and insist that they make well-labelled tables of trials and errors. The final three questions are the most difficult because the geometric problem has to be formulated in numerical terms.

    Discuss and give solutions to most of the problems attempted. **Centre the discussion around the way the errors are used to decide on the next trial.**

2. Conclude with a summary of what makes trial and error powerful - being systematic, recording, and thinking about how to choose the next trial - and explain that it is especially useful when you are faced with an unfamiliar problem that you don't know how to start.

MATHEMATICAL NOTES

NUMBER PUZZLE A

Guess for Number	50	10	15	11	12
One more than twice itself	101	21	31	23	25
Product	5050	210	465	253	300
Error	too big	too small	too big	too small	Hurray!

Note that $-12\tfrac{1}{2}$ is also a solution, not found by this method. Some students may use the fact that the answer is a factor of 300.

NUMBER PUZZLE B

Guess for Number	10	20	15	18	17
Number multiplied by itself	100	400	225	324	289
Total	110	420	240	342	306
Error	too small	too big	too small	too big	just right

Note that -18 is also a solution.

NUMBER PUZZLE C

Number	1	0.5	2	10	20	15
One ÷ number	1	2	0.5	0.1	0.05	0.067
One ÷ twice number	0.5	1	0.25	0.05	0.025	0.033
Total	1.5	3	0.75	0.15	0.075	0.100
Error	too big	even bigger	still too big but better	too big but better	too small	Hurray! (checks with fractions too)

NUMBER PUZZLE D

This problem is included because the positive answer is not a whole number. There are two solutions: +59.5 and -60.

NUMBER PUZZLE E

Answer is $\sqrt[3]{100} = 4.642$. Any desired accuracy can be achieved using trial and error.

OLD CHINESE PROBLEM

There are 14 rabbits and 21 pheasants.

AGES

Rusty is 12 and Wayne is 42.

CHOOKS

The answer (side 5m) is interesting here as an example of the fact that a square encloses the greatest area of all rectangles with a given perimeter.

From the trial and error point of view, the problem is interesting because the function is not monotonic - i.e. the area does not always increase as the side increases.

Width (metres)	1	2	3	4	5	6	7
Length (metres)	9	8	7	6	5	4	3
Area (sq. metres)	9	16	21	24	25	24	21
Trend		bigger	bigger	bigger	bigger	smaller	smaller

Further testing between 4 and 6 supports the idea that the largest possible area is 25 square metres.

MORE CHOOKS

One side 10m, two sides 5m.
The function involved here is also not monotonic, i.e. the area increases until the side reaches 10m, then decreases.

PACKAGES

Dimensions: 1m x 0.25m x 0.25m

Volume: 0.0625 m3 or 62 500 cm3.

Strategies for Problem Solving   Worksheet 8B2-1

TRIAL AND ERROR

NUMBER PUZZLE A

A number when multiplied by one more than twice itself gives a product of 300. What is the number?

NUMBER PUZZLE B

I am thinking of a number. When I multiply my number by itself and then add my number, I get 306. What is my number?

NUMBER PUZZLE C

I was thinking of a number. First I divided one by my number. Then I divided one by twice my number. Then I added the two answers together and got 0.1. What was my number?

NUMBER PUZZLE D

I am thinking of a number. When I multiply my number by itself and then add half my number, I get 3570. What is my number?

NUMBER PUZZLE E

I am thinking of a number. When I multiply it by itself and then by itself again, I get 100. What is my number?

AN OLD CHINESE PROBLEM

In a pen there are rabbits and pheasants. They have between them 35 heads and 98 feet. How many rabbits? How many pheasants?

AGES

Every year of a dog's life is worth seven years of a human's life. If Rusty was a human, she would be twice as old as her owner, Wayne. If Wayne was a dog, he would be six years younger than Rusty. How old is Rusty?

CHOOKS

The Smith family wants to build a chook pen and they have bought enough wire for 19 metres of fence and one gate that is 1m wide. They have decided to make the pen rectangular or a square. What width and length would they make the pen so that the chooks have the largest possible area inside the pen?

MORE CHOOKS

The Jones are also building a chook pen with 19 metres of fencing and a 1 metre gate, but they have decided to use part of their back fence as one side of the chook pen, so they only have to build three sides of a rectangle. Their back fence is 20 metres long. What size should they make their pen?

PACKAGES

To be able to send a package through the post it must measure no more than 1m around (its girth) and it must be no more than 1m long. What is the size of the box with greatest volume that I can send through the post?

Lesson Block 8C

FOUR-CUBE HOUSES

Theme   **A good recording system promotes effective problem solving.**

Aims   To show the benefits of developing **a good system of recording what is done** - especially in parallel with the manipulation of concrete materials.

To practise carrying out **transformations of three dimensional objects.**

To encourage a **systematic approach,** enabling tasks to be divided sensibly amongst group members.

Mathematical Activity

FOUR-CUBE HOUSES

"There is restlessness in the midget town. Some houses are more beautiful than others. Paulus is called in as a trouble shooter. He proposes to rebuild the town. The midgets will live pairwise in houses, each consisting of a drawing room, a kitchen and two bedrooms. All rooms are to be (congruent) cubes, and each house will be built from four cubes, which touch each other along complete faces, thus

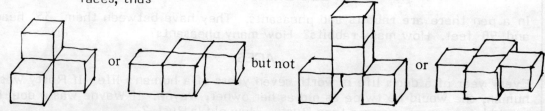

It would be a dull town if all the houses were the same shape".

How many different houses can the midgets build?

Summary of Lesson Block

FOUR-CUBE HOUSES

Lesson 8C1   Doing the FOUR-CUBE HOUSES problem, with class discussion on methods of recording and methods of working systematically.

Lesson 8C2   Using the results found in the first FOUR-CUBE HOUSES lesson. Further visualising of three dimensional shapes is involved.

THREE-IN-A-ROW

Lesson 8C3   A very simple variation of noughts and crosses is analysed. Again the theme is to demonstrate the importance of developing a good system for recording moves.

Materials Required and Teacher Preparation

For Lessons 8C1 and 8C2

(i)   Eight to twelve large cubes are useful for teacher's demonstrations.

(ii)   Each group of 4 children could use up to 60 small cubes. The importance of the cubes depends on the maturity of the children.

(iii) For project 2 (Lesson 8C2) each child may need a copy of Worksheet 8C2-1.

The article "Four-Cube Houses" by Hans Freudenthal in Volume 1 (November 1980) of "For the Learning of Mathematics" describes what happened when this problem was tackled by Dutch third grade children - it is a useful account of how to lead a mathematical investigation for teachers at any level.

For Lesson 8C3, each child could draw a large board and use counters for early experimentation. Soon however, they will need to record moves and individual copies of Worksheet 8C3-1 will then be useful.

SAMPLE LESSON PLAN

Lesson 8C1

1. Refer back briefly to LEAPFROGS, making the point that one of the crucial elements in the solution was to develop some way of recording the moves. Recall some examples of methods used by class members. This problem has a similar feature.

2. Tell your own version of the four-cube houses story. The one quoted as the Mathematical Activity above is taken from Freudenthal's article. Paulus the forest midget is a well known feature on Dutch children's T.V.

   Other settings for the same story can be used. For example, a young architect may be designing houses for a ski village. The rooms have been pre-fabricated as cubic modules.

3. Each student uses cubes to build one or two houses and selected students show theirs to the class using the larger teacher's blocks. Use this activity to make sure everyone understands the rules for building (i.e. cubes must meet along complete faces). Then begin a **discussion of which houses are regarded as the same and which are different,** perhaps by having students indicate if they have made the same house as the one being displayed. Emphasise that houses are the same if the builders' plans are the same - the orientation on the block of land doesn't matter.

   A good example is that the houses in Figures 1a, 1b and 1c are the same, but not the same as the house in Figure 1d.

96 Strategies for Problem Solving

Figure 1a   Figure 1b   Figure 1c   Figure 1d

4. Divide the children into groups and challenge them to **make as many different houses as they can.** (Do not tell them that there are 15 types.) **Suggest they divide the work** amongst themselves in some logical way, so that each group member may take a particular ground plan or look for extensions of a particular three-cube house.

5. Gravity-defying houses (requiring stilts or cantilevering) seem to fascinate children and some will probably want to include these. **Allow this if time permits** after they have found all the ordinary houses. An interesting question to investigate is whether every gravity-defying house is really just an ordinary house placed down on its side or roof.

6. When most groups feel they have found all the houses, begin to collect the complete set from the class. This will prompt more same/different questions and lead naturally to the desire to draw them on the board. As drawing is difficult, a recording system (a notation) needs to be developed. **Discuss possibilities and choose one** - a suggestion is made in the Mathematical Notes below.

7. Students should **record a complete set of houses** in their books.

8. (Optional) Claim that you are not convinced that all the houses have been found. Challenge groups to **demonstrate that they do in fact have them all.** Suggest they use **a systematic approach.** For example they might look in turn at 4, 3, 2, 1 rooms on the ground floor.

9. **Write work up** using the headings "What I discovered" and "What problem-solving idea I found helpful". They should summarise, rather than repeat, what has been done.

Lesson 8C2

The aim of this lesson is to use the results found in the previous lesson and to reinforce the skills of three-dimensional thinking involved. It can follow any number of lines - here are three ideas, but you could follow up a student's suggestion instead.

Project 1: Costing houses

The 15 houses are to be built on a communal block of land. The cost of building a house is $1000 per (unit) square of land covered, plus $1000 per square of external wall, plus $1000 per square of roof. The insides of the houses all cost the same amount. Find the cheapest house and the most expensive house.

Project 2: Elevations

A burglar on a dark night looks at a house from the front and back and the left and the right, seeing

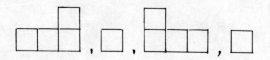

(not necessarily in that order). Which house is it?

The answer is | 1 | 1 | 2 | .

Students could then draw the elevations of any house and challenge each other to find which one it is. Encourage students to make up the elevations mentally then, if necessary, check them physically.

Alternatively Worksheet 8C2-1 could be used.

Project 3: How many five-cube houses can you find?

As there are many more than 15 five-cube houses, it is inappropriate to expect students to prove that they have them all. Every five-cube house can be built by extending a four-cube house, so this is one way to divide up the task. Alternatively, the 23 ground floor plans of the five-cube houses could be used as a basis for the division of labour among group members.

Freudenthal (reference above) reports tremendous activity, with a display of everything from trivial solutions to impressive arithmetical architecture when the third grade children started designing sixty-cube houses.

Lesson 8C3

The aim of this lesson is to further elaborate on the theme (a good recording system promotes efficient problem solving) by using a different type of problem.

1. **Introduce THREE-IN-A-ROW.** It is played like noughts and crosses, but on a different board. The aim is for a player to get three of her own markers in a straight line. The markers can be placed in any of the 7 circles on the board.

    The task in this lesson is to find a winning strategy, i.e. an unbeatable strategy for either the first or the second player. It might be introduced as follows: For the school open day, the computers will be programmed to play this game against the visitors. The programmer must be given clear instructions about the game so he can make the computer always win.

    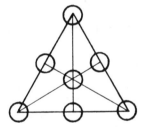

2. Let students **get the feel of the game** by playing it in pairs. Initially students may use counters on large boards, but it is better if they draw their own boards or use Worksheet 8C3-1. When they begin conjecturing as to who can win, make sure the task is made clear to them and they understand that they will have to **write down instructions covering all eventualities** that will lead to a win for their chosen player. Only one way of winning is required, but it must take into account all possible moves by the opposition.

3. During the lesson, emphasise the value of **looking at symmetry** and the **necessity for keeping a written record** in a convenient notation.

4. Ask students to write down their instructions clearly. Before they do this, **discuss some possible systems of notation** that they can use.

5. Students who finish quickly could analyse conventional noughts and crosses in the same way. (Played 'properly', this always ends in a draw.)

MATHEMATICAL NOTES

FOUR-CUBE HOUSES

Finding all the different houses

Note that two houses are the same if they can be transformed from one to the other by a rotation about a vertical axis.

Drawing the houses is too difficult, so an efficient system of recording them is required. One good system is to draw the ground floor plan and write how many cubes are standing above each square.

Using this notation, the fifteen four-cube houses are

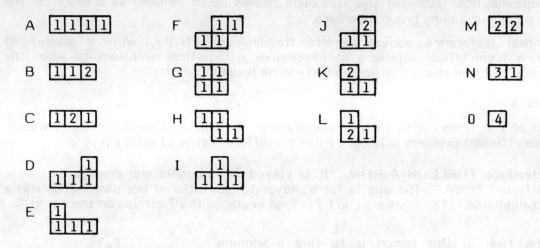

Project 1: Costing Houses

The costs have been chosen so that the difference in price depends on the surface area of the house. All the houses cost the same amount except houses G and M above, which are $2000 cheaper. Most students use direct calculation, but some students will notice that the price will be the same for all houses that are geometrically congruent (under any rotations or reflections). Others will see that the cost of a house is quickly found from the number of internal walls. For example, the four individual cubes that make up house L have 24 faces, but 6 of these form the internal walls and floor. Thus the cost of the house is $24000 - $6000 = $18000.

Project 2: Elevations

There are nine different sets of elevations. The houses A, B, C, G, M N, O are distinguished by their sets of elevations. The three houses J, K and L share a set of elevations, as do the remaining five houses D, E, F, H, I.

The answers to Worksheet 8C2-1 are: house 1 is A, house 2 is M, house 3 is N, house 4 is J or K or L, house 5 is C, and house 6 is D, E, F, H or I.

Project 3: How Many Five-Cube Houses Can You Find?

The 23 ground floor plans are one single cube, two adjacent cubes, three cubes in a row or L-shaped, the seven one-storey four-cube houses and the twelve pentominoes.

THREE-IN-A-ROW

The first player can always win.

Let Ai denote the ith move of 1st player and let Bj denote the jth move of second player.

A Winning Strategy

Place A1 in the centre. If B1 is at a corner, place A2 in another corner. This forces player B to block with B2 on the side (otherwise A wins). A3 can then be at the remaining corner, opening up two chances for A, as in Figure 1. If B1 is on the side, the play is as in Figure 2.

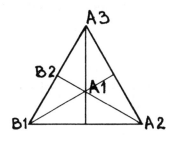

Figure 1

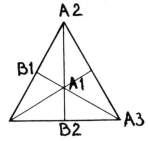

Figure 2

Although only one winning strategy is needed to ensure a win every time, there are other winning strategies. For example, if A1 is placed at a corner, there are four essentially different positions for B1 and the first player is still guaranteed a win, as shown below.

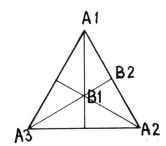 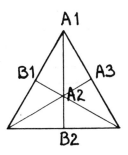 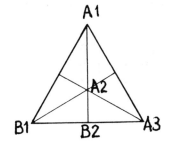 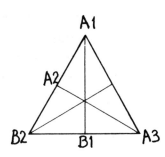

Strategies for Problem Solving   Worksheet 8C2-1

FOUR-CUBE HOUSES FROM DIFFERENT POINTS OF VIEW

Here are the ELEVATIONS of six four-cube houses. The views of the house from the front and back, right side and left side are shown - but they are not necessarily in that order. Draw the houses in the space provided.

House 1

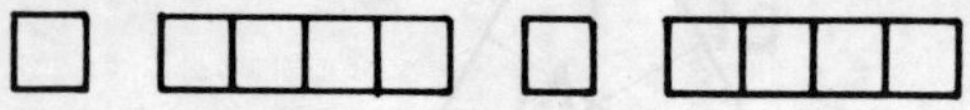

House 4

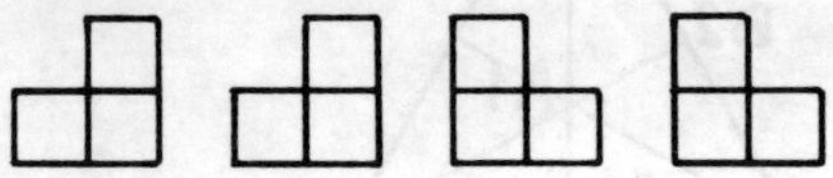

House 2

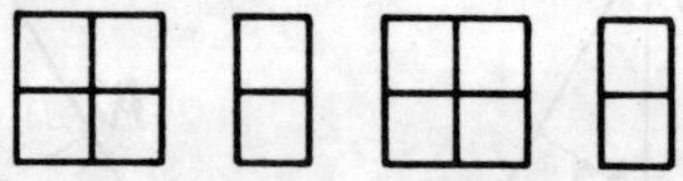

House 5

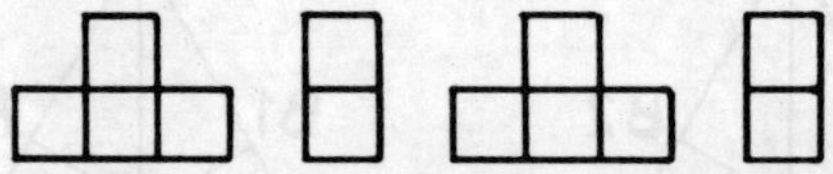

House 3

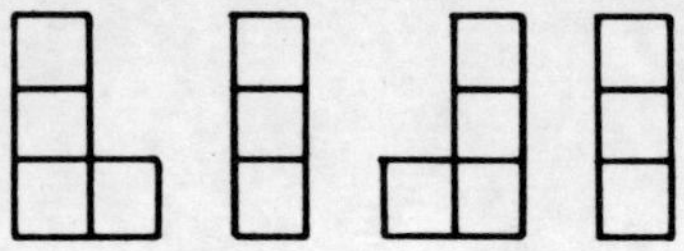

House 6

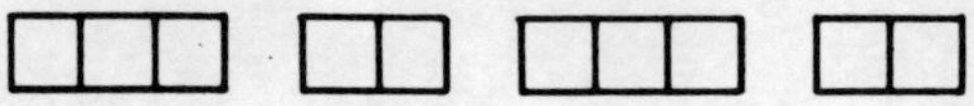

Strategies for Problem Solving   Worksheet 8C3-1

THREE-IN-A-ROW

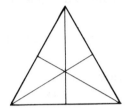   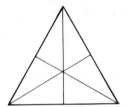   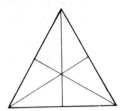

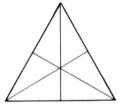   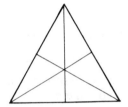   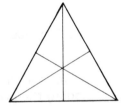

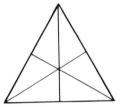   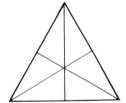   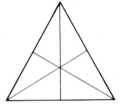

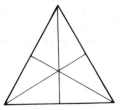 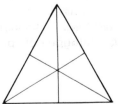 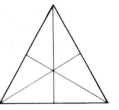

Lesson Block 8D

FIBONACCI NUMBERS

Theme   Look for a simpler problem.

Aims   To demonstrate the strategy "look for a simpler problem". This can be a very sophisticated strategy but it is introduced here in a limited context where patterns are very obvious.

To introduce children to some of the interesting properties of the Fibonacci sequence.

To develop an appreciation of ways of testing patterns and (for more able students) some skills in explaining why a pattern is valid.

Mathematical Activity

FIBONACCI NUMBERS

The Fibonacci sequence of numbers is

$$1, 1, 2, 3, 5, 8, 13, 21, \ldots .$$

After the initial terms, each term is constructed by adding the previous two terms.

The sequence has many remarkable properties and relates to a variety of natural situations including the ancestry of male bees, the number of segments in spirals on pine cones and pineapples and the number of petals on some flowers.

Summary of Lesson Block

Lesson 8D1   An introduction to the Fibonacci numbers and to some of their properties. The "look for a simpler problem" strategy is used to save calculations - see Worksheet 8D1-1.

Lesson 8D2   The problems on Worksheet 8D2-1 introduce children to other occurrences of the Fibonacci sequence, where the properties previously discovered can be applied. Worksheet 8D2-2 is for fast workers and looks at some of the analogous properties of the Lucas numbers.

Material Required and Teacher Preparation.

Before Lesson 8D1, teachers could consult the reference below, or others, for more details about the natural occurences of the Fibonacci sequence. Some information is given in the Mathematical Notes below. Individual copies of the worksheets may be used. Calculators are useful.

Reference: V.E. Hoggatt **"Fibonacci and Lucas Numbers"**, Houghton Mifflin, New York, 1969.

SAMPLE LESSON PLAN

Lesson 8D1

1. Introduce the lesson, perhaps by explaining how **mathematicians always like to do things the easy way.** Rather than do a lot of hard calculations, it is often best to solve easier problems first and use a pattern to find the answer to the hard problem. Label this strategy "look for a simpler problem" and write this on the board - **children learn more readily if strategies are labelled and discussed explicitly.**

2. Introduce the Fibonacci sequence, perhaps using the rabbit story outlined in the Mathematical Notes. Other historical and biological details are also given there.

   Describe how to find the terms of the sequence. Make sure everyone understands the terminology "3rd term, 9th term" etc.

   List the first few Fibonacci numbers on the board and **ask students to calculate up to about the fifteenth term.** Write them down for later reference (on the board and in the students' books)

   1, 1, 2, 3, 5, 8, 13, 21, 34, 55, 89, 144, 233, 377, 610, 987, ...

3. Hand out Worksheet 8D1-1 and illustrate the strategy with Question 1 from Worksheet 8D1-1.

   > "What is the result of adding every second Fibonacci number, starting at the first and ending with the twenty-ninth?"
   > (The answer is the thirtieth Fibonacci number.)

   Discuss the two options: you can either **do the calculation** as it is OR you can **look at simpler problems** (in this case adding less numbers) to find a pattern that could be used.

   Complete the table as on Worksheet 8D1-1 and discuss with the class any patterns observed.

Table	
Simpler Problems	Answers
1	1
1+2	3
1+2+5	8
1+2+5+13	21
1+2+5+13+34	55

104 Strategies for Problem Solving

Emphasise the need to test the pattern by predicting results that can easily be checked. A chalkboard summary of the steps involved would be useful.

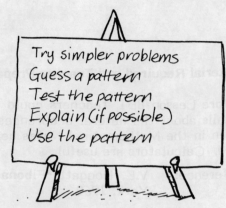

Guess a pattern: Sum is the next Fibonacci number.

Test the pattern: 1+2+5+13+34+89 = 144 and perhaps do the next case as well.

Use the pattern to find the answer: Sum is 30th Fibonacci number.

Explain why: Some children will be able to offer an outline of an inductive argument to show what is going on, along these lines:

When we know   1+2+5+13   = 21,
then   1+2+5+13+34 = 21+34
                                          = sum of two consecutive Fibonacci numbers
                                          = the next Fibonacci number.

This happens in every case.

5. Compare the two options referred to above. Direct calculation may sometimes have been a little quicker, but finding and using the pattern gives valuable knowledge about the sequence that can be used in other circumstances.

6. Allow students to work on Questions 2 and 3 from Worksheet 8D1-1. They both follow on very similar lines to the example above.

Lesson 8D2

1. In this lesson children tackle the problem situations from Worksheet 8D2-1. All of these problems involve the Fibonacci sequence, although children should not be told this at the beginning. **By looking at simpler cases** (fewer steps, less money, shorter path) **the Fibonacci pattern emerges.**

2. When testing a pattern to see if it is valid, some children tend to make predictions about cases that are impossible (or extremely difficult) to check independently. For example, when testing the pattern observed in Question 1 (Worksheet 8D1-1) some children will try to test it on the sum of twenty Fibonacci numbers, where an independent check by direct calculation is impossible. Watch for this and explain the logical defect where it occurs. Checking has to be cost efficient - there is no point having to do calculations as hard as the original question!

3. Conclude with a **brief resume of how examining the simpler problems has helped.**

4. More able children can tackle Worksheet 8D2-2, which deals with the Lucas numbers, close relatives of the Fibonacci numbers.

MATHEMATICAL NOTES

HISTORICAL BACKGROUND

Leonardo Fibonacci was born in Pisa, Italy, and because of that he was also known as Leonardo Pisano, or Leonardo of Pisa. While his father worked on the northern coast of Africa, Fibonacci had a Moorish schoolmaster who introduced him to the Hindu-Arabic numeration system that we use today.

After widespread travel and extensive study of computational systems, Fibonacci wrote, in 1202, the **"Liber Abaci"** in which he explained the Hindu-Arabic numerals and how much easier they were to use than Roman numerals. The famous problem of the rabbits illustrated how easily calculations could be done with the new numerals. Fibonacci introduced the rabbit problem in **"Liber Abaci"** through a story along these lines:

Suppose a pair of rabbits produce one pair of offspring when they are exactly two months old and one pair each month thereafter. A pair of new born rabbits is put into a cage on January 1st. How many rabbits will there be on January 1st of the next year? (Of course, rabbits don't die or do anything else unforseen.)

Teachers are advised to work through this situation in detail, listing separately the numbers of pairs of mature and immature rabbits in a table.

Month	Jan	Feb	Mar	Apr	May	Jun	Jul	Aug
Number of pairs of rabbits	1	1	2	3	5	8	13	21

The Fibonacci sequence arises because the number of pairs of rabbits in the cage in a given month is equal to the number there in the previous month **plus** the number of pairs of babies born, and this is just equal to the number of mature pairs present, i.e. all those rabbits in the cage two months previously. For example, in September there will be the 21 pairs present in August **plus** 13 pairs of babies born to the 13 mature pairs (they were all those present in July).

Number of rabbits at month $(n + 2)$ = number of rabbits at month $(n + 1)$ + number of rabbits at month n

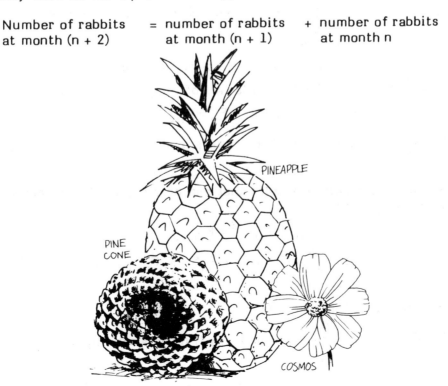

PINEAPPLE

PINE CONE

COSMOS

Fibonacci numbers also occur in a variety of other biological situations. The number of petals (or petal-like parts) of a flower such as the Aster, Cosmos or Daisy in the composite family is consistently a Fibonacci number (21, 34, 55 or 89). The number of spirals in the seed patterns of sunflowers, scale patterns of pinecones and pineapples are usually Fibonacci numbers. A male bee has a mother but no father (it is born from an unfertilized egg) but a female bee has both a mother and a father. The number of ancestors of a male bee at any particular generation is a Fibonacci number.

Number of male bees	Number of female bees	Number of bees
1	0	1
0	1	1
1	1	2
1	2	3
2	3	5
3	5	8

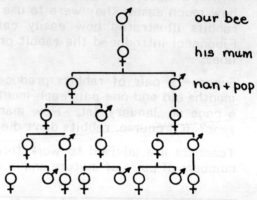

A MALE BEE'S FAMILY TREE

Legend: ♂ male bee
♀ female bee

SOLUTIONS TO WORKSHEET 8D1-1

Let F_n denote the nth Fibonacci number.

Question 1: The answer is the thirtieth Fibonacci number. This is part of the general pattern:

$$F_1 + F_3 + F_5 + \ldots + F_{2n-1} = F_{2n}.$$

Question 2: The answer is one less than the thirty-first number,

i.e. $(514229 + 832040) - 1 = 1346268$. This is part of the general pattern:

$$F_2 + F_4 + \ldots + F_{2n} = F_{2n+1} - 1.$$

Question 3: The answer is one less than the thirty-second number,

ie. $(832040 + 1346269) - 1 = 2178308$. This is part of the general pattern:

$$F_1 + F_2 + F_3 + \ldots + F_n = F_{n+2} - 1.$$

This pattern is a simple consequence of the results found in questions 1 and 2 above.

There are **many other interesting results** that can be similarly found through guided discovery. Two easy ones are

$$F_n^2 + F_{n+1}^2 = F_{2n+1}$$

and $F_1^2 + F_2^2 + \ldots + F_n^2 = F_n F_{n+1}.$

One of the nicest results is this theorem:

F_n is divisible by F_m if and only if n is divisible by m.

SOLUTIONS TO WORKSHEET 8D2-1

STEPS It's the Fibonacci sequence again! Why?

To go n+2 steps, I can either go up the first n+1 steps in any way and then the final step OR go up the first n steps in any way and then the final two together.

Do n+1 steps any way and then the final step   OR   do n steps in any way and then take the final two steps.

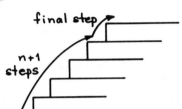 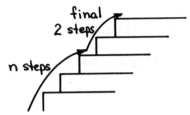

GARDEN PATH Order matters again, and the Fibonacci sequence applies.

When I tile a path of length (n+2)x2 with 1x2 tiles, the final tile(s) can go

like this   or like this

In this case we have had to tile the path of length n+1.

In this case we have had to tile a path of length n.

∴ number of ways to tile a path of length n+2 = number of ways for n + 1 + number of ways for n

SMALL CHANGE This would be the Fibonacci sequence if the order in which the coins were handed over mattered. As it is, there are n+1 ways of paying 20xn cents, namely n twenties, n-1 twenties plus 2 tens, n-2 twenties plus ......, all tens.

To pay 20xn + 10 cents, you must pay an extra 10 cents and then 20xn cents in any of the n+1 ways shown above.

∴ 20n cents can be paid in n+1 ways
and 20n + 10 cents can be paid in n+1 ways.

LUCAS NUMBERS

Q1: 1, 3, 4, 7, 11, 18, 29, 47, 76, 123, 199, 322, 521, 843, 1364.

Q2: $L_1 + L_2 + \ldots + L_n = L_{n+2} - 3$

Q3: $L_1 + L_3 + \ldots + L_{2n-1} = L_{2n} - 2$

Q4: $L_2 + L_4 + \ldots + L_{2n} = L_{2n+1} - 1$

} Similar explanations to corresponding Fibonacci results.

Q5: Every third Lucas number is even.

Q6: $L_1^2 + L_2^2 + \ldots + L_n^2 = L_n L_{n+1} - 2$ (too hard to prove).

Strategies for Problem Solving   Worksheet 8D1-1

FIBONACCI NUMBERS

The Fibonacci numbers are 1, 1, 2, 3, 5, 8, 13, 21, 34, ...

When you try the questions below, you will find it helpful to know that the 29th Fibonacci number is 514229 and the 30th is 832040.

Question 1: What is the result of adding every second Fibonacci number, starting at the first and ending with the twenty ninth?

Simpler Problems	Answers
1	= 1
1+2	= 3
1+2+5	= 8

Question 2: What is the result of adding every second Fibonacci number starting at the second number and ending at the thirtieth number?

Simpler Problems	Answers
1	= 1
1+3	= 4
1+3+8	= 12

Question 3: What is the sum of the first thirty Fibonacci numbers?

Simpler Problems	Answers

Strategies for Problem Solving   Worksheet 8D2-1

LOOK AT SIMPLER PROBLEMS

STEPS

There are four steps to my front door. I like to go up taking either one step or two steps at a time.

There are 5 different ways that I can go up.

Carolyn has five steps to her front door and Joshua has six. **In how many ways can they climb their steps?**

Can you find a way of predicting the answer for other numbers of steps?

BEES

A male bee has a mother but no father. A female bee has both a mother and a father. How many great, great, great, grand parents does a male bee have? What about humans?

GARDEN PATH

I want to pave a garden path that is 60 cm wide and 210 cm long, with pavers that measure 30 cm by 60 cm. **How many pavers will I need? In how many different ways can I lay the pavers?**

One example:

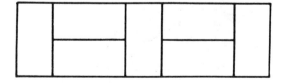

SMALL CHANGE

In my purse I have lots of 10 cent pieces and 20 cent pieces. If I buy a cake for 50 cents there are three different ways that I can pay the exact money:

- five 10 cent pieces
- three 10 cent pieces and one 20 cent
- one 10 cent piece and two 20 cents.

In how many different ways could I give exact money for a $1.10 drink? Can you work out a rule for predicting the number of different ways I can pay bills of other sizes?

Strategies for Problem Solving   Worksheet 8D 2-2

LUCAS NUMBERS

Lucas numbers are close relatives of the Fibonacci numbers, having many analogous properties and similar biological connections.

They are constructed in the same way as the Fibonacci numbers, but the starting numbers are different. The first Lucas numbers are 1, 3, 4, 7, 11, 18, 29, 47.... The 29th Lucas number is 1149851 and the 30th is 1860498.

1. Write down the first fifteen Lucas numbers.

2. Find the sum of the first thirty Lucas numbers.

3. Find the sum of every second Lucas number, starting at the first and ending at the 29th.

4. Find the sum of every second Lucas number, starting at the second and ending at the 30th.

5. Which Lucas numbers are even?

6. Square the first 10 Lucas numbers and add up all the squares. What is the sum? (This pattern is harder to spot!)

FIND OUT HOW FIBONACCI AND LUCAS NUMBERS ARISE IN NATURE

Lesson Block 8E

CRYPTARITHMS

Theme   **Working systematically.**

Aims   To demonstrate the benefits of using a **systematic approach.**

              To demonstrate the necessity of **keeping a record** of trials made.

              To encourage **logical analysis.**

              To encourage **planning ahead** in problem solving.

Mathematical Activity

CRYPTARITHMS

```
  TWO
 +TWO
 ----
 FOUR
```

This is a cryptarithm - a calculation where the digits have been replaced by letters. The aim is to find out what the numbers are.

Summary of Lesson Block

Lesson 8E1   Introduce cryptarithms. Teacher and class solve some easy ones to demonstrate techniques. Children work in small teams on one or more harder cryptarithms.

Materials Required and Teacher Preparation

Copies of Worksheet 8E1-1. Calculators and multiplication tables may speed up the trial and error process.

SAMPLE LESSON PLAN

Lesson 8E1

1. Hand out Worksheet 8E1-1. Explain what cryptarithms are and the rules ("clues") for solving them, as given on the worksheet.

2. Write on the top of the board the helpful hints WORK SYSTEMATICALLY and WRITE DOWN WHAT YOU DO. Leave these on the board throughout the lesson and add any others that arise during discussion.

112 Strategies for Problem Solving

3. Begin with some easy cryptarithms (examples 1-4 from Worksheet 8E1-1) to demonstrate some of the techniques. Allow students a few minutes for trial and error before showing a more systematic method (or having pupils suggest it).

Examples

| 1. | A
xA
―――
BA | 2. | A
xA
―――
AB | 3. | AB
+BA
―――
CAC | 4. | AB
xAB
―――
ACC |

Solutions

1. A = 5, B = 2
   or A = 6, B = 3
2. None
3. A = 2
   B = 9
   C = 1
4. A = 1
   B = 2
   C = 4

Examples 1 and 2 can be solved readily by testing the nine possibilities for A (since A ≠ 0). Example 2 is the only one on the Worksheet with no solutions.

For example 3, point out three important features of the analysis:
- (i) Two numbers less than 100 must add to less than 200, so C = 1. Looking at the units column, we see A + B = 1 or A + B = 11.
- (ii) Since neither A nor B can be zero, A + B = 11. (Always watch out for carrying!).
- (iii) The solution can now be found quickly and easily by trial and error.

For example 4, as usual, since A is the first digit, A ≠ 0. Also, as in example 3, two numbers less than 100 must add to less than 200, so A = 1. The nine possibilities for B give one solution: A = 1, B = 2, C = 4.

4. Move on to more challenging problems - if necessary asking students to finish one off at home. All students should try example 5: TWO
                                                                                  +TWO
                                                                                  ―――
   This one is easy because there are many solutions.   FOUR
   Discuss it later in class. Encourage students to work in teams and to divide the various possibilities amongst team members.

5. High fliers should try at least one of examples 6-8. (See the Mathematical Notes for solutions.)

   Do not attempt to demonstrate how to find solutions to the class as a whole, but offer hints when appropriate.

6. End the lesson by referring back to the helpful hints on the board and summarising how they have been used. Students should **write up** one harder cryptarithm. They need not repeat all the details, but should **look back** over the calculation, **simplify** it as much as possible and then **summarise it**.

MATHEMATICAL NOTES

Example 5

```
  TWO
 +TWO
 ----
 FOUR
```

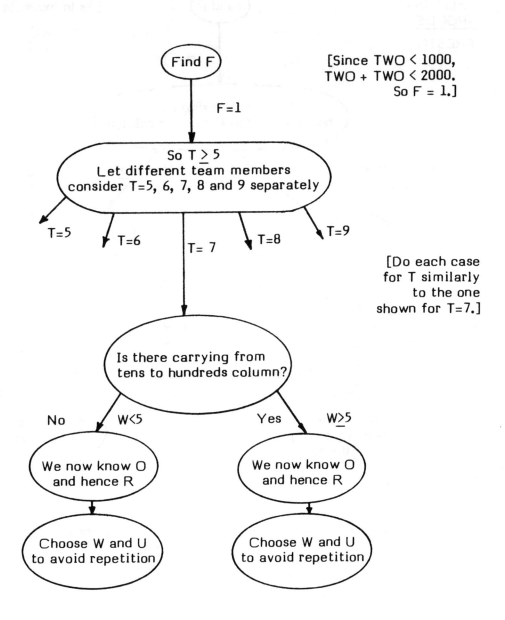

Solutions

```
   734      765      836      867      928      938
 + 734    + 765    + 836    + 867    + 928    + 938
  ----     ----     ----     ----     ----     ----
  1468     1530     1672     1734     1856     1876
```

114 Strategies for Problem Solving

Example 6

HOCUS
+POCUS
─────
PRESTO

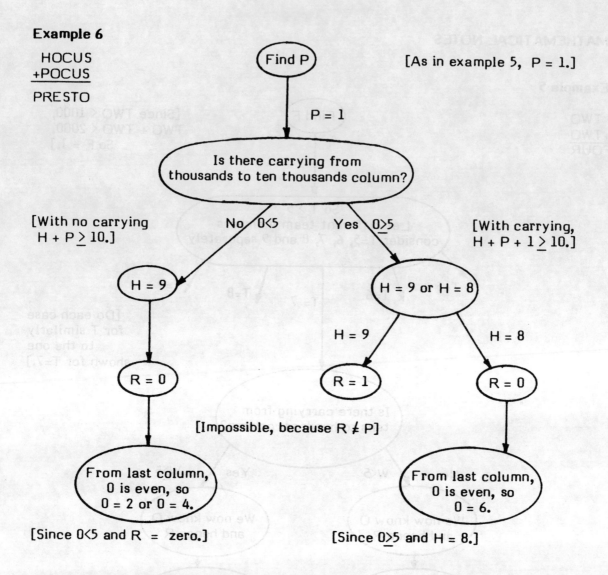

Different team members can now consider these two situations to get the unique **solution**

```
  92 836
+ 12 836
────────
 105 672
```

Example 7

```
  SEND
+ MORE
-------
 MONEY
```

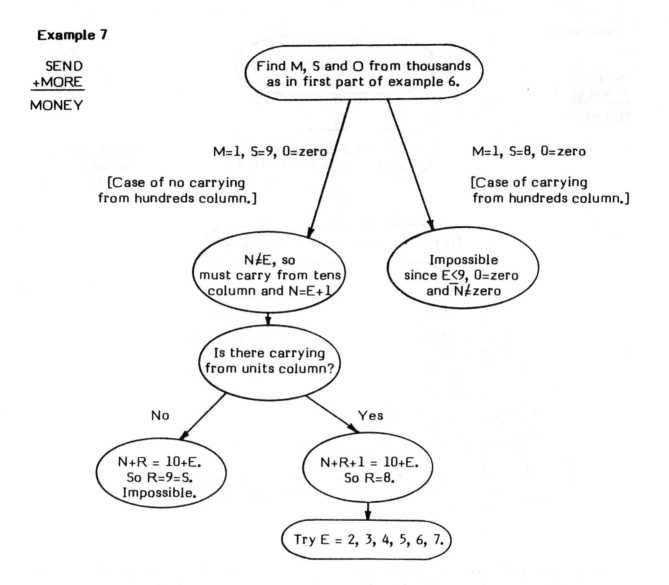

Solution

```
   9 567
+  1 085
--------
  10 652
```

Example 8

```
    A
MERRY
 XMAS
TURKEY
```

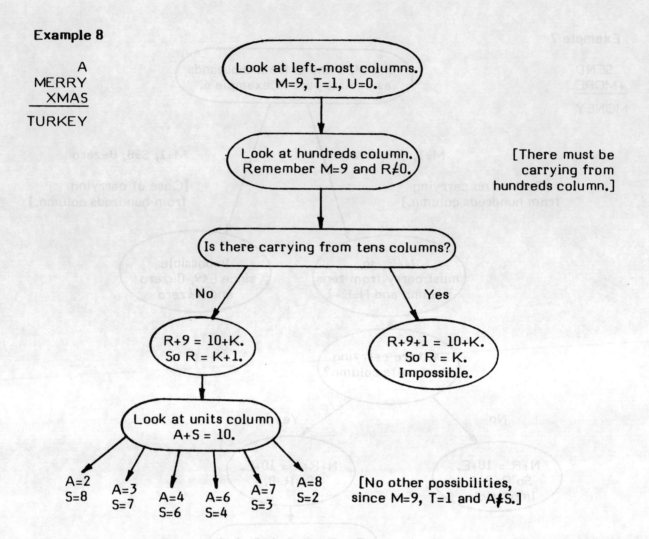

Can now divide into teams and use trial and error on the information above (M=9, T=1, U=0 and R=K+1) and the additional information that 1+R+A=E (from tens column) and E+X+1 = R+10 (thousands column).

[It helps to leave Y to last!]

Solution

```
       2
  97 445
   6 928
 -------
 104 375
```

Strategies for Problem Solving   Worksheet 8E1-1

CRYPTARITHMS

Cryptarithms are puzzles where numbers have been replaced by letters.
Here are some clues to help you solve the puzzles
.   each letter in a cryptarithm is used for only one digit.
.   different letters stand for different digits (repeating is not allowed).
.   the letter O may or may not stand for the number 0 (zero).
.   in a number like AB, the first digit cannot be zero (so A≠0 here).

```
1.    A         2.    A         3.   AB         4.   AB
     xA              xA             xBA             xAB
     ──              ──             ───             ───
     BA              AB             CAC             CAC
```

```
5.   TWO            6.   SEND
    +TWO               +MORE
    ────               ─────
    FOUR               MONEY
```

```
7.   HOCUS
    +POCUS
    ──────
    PRESTO
```

```
8.        A
       MERRY
       XMAS
      ──────
      TURKEY
```

Lesson Block 8F

CARD TRICKS

Theme Explaining why.

Aims To practise finding explanations.

To show how mathematics can explain some magic and mystery.

Mathematical Activity

FIFTEEN CARDS

Deal 15 cards from the pack into 3 piles of 5, face up. The person watching secretly chooses one card. Ask the person which pile the card is in. Pick up all 3 piles, making sure that the pile containing the secret card is in the middle of the other two.

Twice more deal the cards into 3 piles, ask the person which pile it is in and pick the piles up in the same way.

Now deal the cards face up slowly, looking carefully at each one.

The secret card will be the eighth card from the top.

Explain why this trick works.

Summary of Lesson Block

Lesson 8F1 Teacher demonstrates card tricks and students learn how to perform one trick each and then try to explain how and why it works.

Lesson 8F2 Students learn other tricks and think about why they work.

Materials Required and Teacher Preparation

Teachers should thoroughly **learn how to perform each trick.** They are more entertaining if performed confidently and with mystery.

Each student will need a **copy of one of the Worksheets 8F1-1, 8F1-2, 8F1-3 or 8F1-4.** All four tricks need not be used, but at least 2 must be used.

At least one pack of cards is needed for every pair of students. More tricks can be found in **"Mathematics, Magic and Mystery"** Martin Gardner (Dover, 1956, New York).

SAMPLE LESSON PLAN

Lesson 8F1

1. Begin the lesson by making some outrageous claims about your powers of magic and mind reading and **perform two or more of the card tricks.** Your performance will help students interpret instructions later, as well as entertain them.

2. Discuss, after the performance, whether the tricks are in fact due to your amazing magical powers or to some more "scientific" process.

3. Give each pair of students one of the Worksheets with the instructions for performing a trick and help them, where necessary, to learn how to do it. It is best for students to work in pairs as they will need partners to practise with. Explain that the tricks will be more convincing if they are performed confidently and with showmanship.

4. Each student performs his or her trick for one or more other students who have learnt other tricks.

5. After these performances, students should consider why the trick works and **write down as clear an explanation as possible.** Several drafts will be necessary.

Lesson 8F2

1. Each student could be given other tricks to learn, perform and then explain.

MATHEMATICAL NOTES

THE SECRET CARD

In this case, it is fairly easy to see that, on the second deal, column i has become row i and the secret card has been uniquely indentified by its row and column.

FIFTEEN CARDS

When the cards are picked up after the first deal, imagine that the cards are numbered from 1 to 15. Since the pile with the secret card was placed in the middle, it is amongst the "middle five" (i.e. cards 6 to 10). So the second deal looks like this:

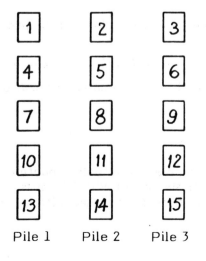

120   Strategies for Problem Solving

By placing the pile with the secret card in the middle again, the third deal will look like one of the following.

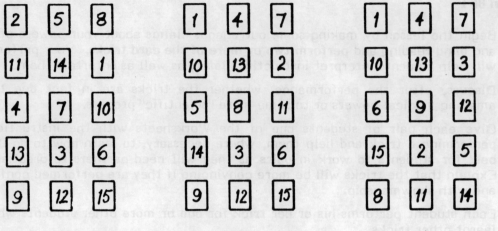

Secret card in pile 1   Secret card in pile 2   Secret card in pile 3
after second deal   after second deal   after second deal
- it must have been 7 or 10.   - it must have been 8.   - it must have been 6 or 9.

In each case, placing the pile containing the secret card in the middle results in the secret card being in the exact middle of the pile - the eighth card.

THE ACE HOTEL

After putting the four piles on top of one another the pack looks like

    A J Q K   A J Q K   A J Q K   A J Q K

    1st suit   2nd suit   3rd suit   4th suit

You can imagine the pack as being circular - e.g.

```
              A
          K       J
        Q           Q
      J               K
      A               A
      K               J
        Q           Q
          J       K
              A
```

No matter where you cut the pack, this order will be preserved, so you will end up with something like

    Q K A J Q K A J Q K A J Q K A J

If you cut between a J and Q, for example, this will give Q K A J as the piles when you deal.

COUNT YOUR CARDS

A very sneaky trick! The magician is actually counting **cards in the pile she has taken.** So, not surprisingly, if she starts with 20 cards, puts 3 aside at some stage and **continues counting from where she finished with the spectator,** she will end up at 17.

Strategies for Problem Solving   Worksheet 8F1-1

THE SECRET CARD

How to perform the trick.

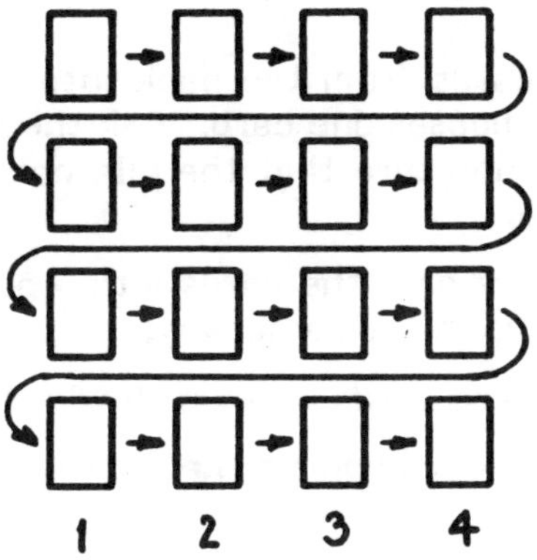

Explain that you are going to demonstrate your special mindreading powers. Deal 16 cards from the pack in 4 rows of 4 columns, face upwards. Be sure to deal the cards "horizontally" row by row as shown.

Ask someone in the audience to secretly select a card, only telling you which of the **vertical columns** the card is in. Now pick up the cards **column by column** putting column 1 on top, then column 2, then 3, then 4.

Deal the cards again, exactly the same way as before. Again ask the person which **vertical column** the card is in.

Now you can **show the person the secret card.**

How to find the secret card.

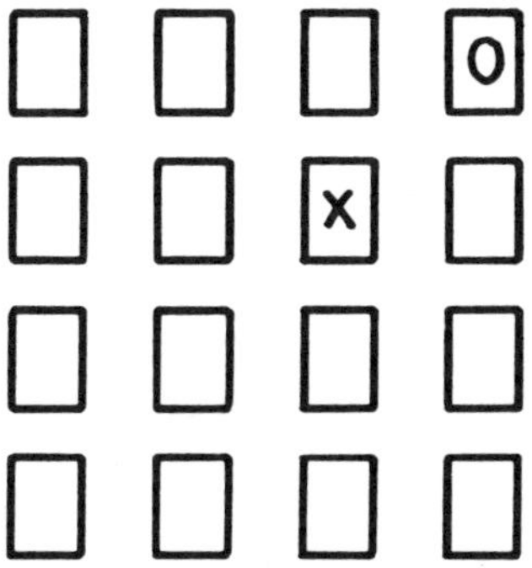

If the card was at first in column 1, it will now be in the first (top) row. You have just been told which column it is in, so you know which is the secret card. Similarly, if it was in column 2, it is now in row 2 and so on.

For example, if the card started off in column 2 and is now in column 3, it will be the one marked X in the diagram.

If it started in column 1 and is now in column 4, it is the one marked O.

Explain why this trick works.

Hint: To help you work out why this trick works, try it with diamonds in column 1, hearts in column 2, spades in 3, and clubs in 4. Watch how the columns are turned into rows when you deal them out for the second time.

Strategies for Problem Solving   Worksheet 8F1-2

FIFTEEN CARDS

How to perform the trick.

Deal 15 cards from the pack into 3 piles of 5, face up. The person watching secretly chooses one card. Ask the person which pile the card is in. Pick up all 3 piles, making sure that the pile containing the secret card is in the middle of the other two.

Twice more deal the cards into 3 piles, ask the person which pile it is in and pick the piles up in the same way.

Now deal the cards face up slowly, looking carefully at each one.

The secret card will be the eighth card from the top.

Explain why this trick works.

Hint: Imagine the secret card was the first card dealt out. Where would it be after the second and third deals? Now imagine the secret card was the second one and trace where it goes. Continue like this.

A Variation.

Deal 21 cards into 3 piles of 7 four times.

The secret card is then the eleventh from the top.

Can you make up other variations?

Strategies for Problem Solving   Worksheet 8F1-3

THE ACE HOTEL

How to perform the trick.

Take the picture cards (aces, jacks, kings, queens) out of the pack.
Tell this story as you put out the cards.

"A hotel had 4 rooms."

Put out the 4 aces.

"Four boys booked into the rooms."

Put the jacks on the aces, following suit.

"They were followed by their mothers."

Put the queens on the jacks.

"They were followed by their fathers."

Put the kings on the queens.

"Then they had an argument and decided to separate."

Put the 4 piles on top of each other and cut the pack – i.e. lift up part of the pack and put it at the bottom.

Cut the pack seven more times. Then deal the cards, face up, into 4 piles.

There should be one pile of aces, one of jacks, one of queens and one of kings.

Explain why this trick works.

Hint: Does it matter if you cut the pack less than 8 times? What if you don't cut it at all?

Strategies for Problem Solving   Worksheet 8F1-4

COUNT YOUR CARDS

How to perform the trick.

The magician begins by explaining that she can tell how many cards there are in the spectator's pile, just by looking at the pile. She then asks the spectator to take some cards from the pack - it must be less than half the pack. The magician takes a larger pile from the pack and quickly counts them (secretly, if possible, but in any case without telling anyone the result).

The rest of the trick depends on the number of cards in the magician's pile. Let's pretend there are 20 cards in the magician's pile. She then says: **"I have as many cards as you have, plus 3 cards and enough left to make 17. Please put out your cards slowly, one by one, counting them out loud."**

While the spectator puts out his cards one by one, counting out loud, the magician puts out the same number of cards, one by one, from her pile. Then, in accordance with her statement, she puts 3 aside. She then continues counting her cards from where the spectator stopped. The last number will be 17.

How to choose your "magic" numbers.

Notice that $20 = 3 + 17$. In this case, where there are 20 cards in the magician's pile, she could choose different numbers adding up to 20. For example, she could say: "I have as many cards as you have, plus 2 cards and enough left to make 18." But you need to be careful not to make the number of cards put aside too big.

If the magician had 26 cards in her pile, she could put 3 aside and count on to 23 (since $26 = 3 + 23$).

The trick is most convincing if it is done several times with different numbers.

Explain why this trick works.

Hint: Think about whose cards the magician is really counting.

Lesson Block 8G

STAIRCASE NUMBERS

Theme An investigation using strategies previously learnt.

Aims To revise, and use in combination, strategies such as

- looking for and using patterns
- working systematically
- recording what you do.

To learn some techniques for **explaining what you have done so someone else can understand.**

Mathematical Activity

STAIRCASE NUMBERS

A staircase number is a number which can be expressed as a sum of consecutive numbers. For example 10, 7 and 12 are staircase numbers because

$10 = 1 + 2 + 3 + 4,$ $7 = 3 + 4,$ $12 = 3 + 4 + 5.$

We can imagine the "staircases"

 10   7   12

In the case of 10, the heights of the **stairs** are 1, 2, 3, and 4.

The number 4 is not a staircase number as the only way of writing it as a sum of consecutive numbers is the trivial and uninteresting way with one stair, of height 4.

Which numbers are staircase numbers and which are not?

Find a recipe for writing a number as a sum of consecutive numbers.

126 Strategies for Problem Solving

Summary of Lesson Block

Lesson 8G1   Introduce staircase numbers and begin investigation.

Lesson 8G2   Continue the investigation.

Lesson 8G3   Write a report explaining what has been discovered and, where possible, showing why it is correct.

Materials Required and Teacher Preparation

The Mathematical Notes discuss ways in which the children might proceed with their investigation. It is important that teachers **try the problem themselves,** avoiding the use of algebra or formulae for arithmetic progressions.

If the teacher intends to introduce the problem concretely, a supply of **small cubes** or **cuisenaire rods** or equivalent are needed for Lesson 8G1.

The teacher needs to **decide on the format of reports** and provide any necessary materials for Lesson 8G3.

SAMPLE LESSON PLAN

Lesson 8G1

1. **Define staircase numbers** using either the abstract approach (a sum of consecutive integers, eg. $10 = 1 + 2 + 3 + 4$) or a concrete approach using small cubes or cuisenaire rods or similar aids to build staircases. (If using rods make it clear that the rods are not limited in size for the problem.)

2. With the whole class, **look at some simple examples of staircase numbers.**

    (i) Is 18 a staircase number? If so, what are the stairs?

    [$18 = 5 + 6 + 7$ and also $18 = 3 + 4 + 5 + 6$]

    (ii) Is 4 a staircase number?
    [No! Experimenting with 4 cubes shows that 4 is not a staircase number, unless we accept a staircase with only one stair. If we do accept this, then all numbers are staircase numbers and the problem becomes much less interesting.]

    (iii) Is 2001 a staircase number?
    [Yes, since $2001 = 1000 + 1001 = 666 + 667 + 668$. There are seven ways altogether.]

    Encourage, during these examples, any **preliminary conjectures** students may wish to make. [For example, all odd numbers are easily done and 18 can be done by taking 3 stairs around the middle number 6, which is 1/3 of 18].

3. **Explain the task** and that the results will be **written up very carefully** as a special report, if possible for display at an important school event. Summarise the task on the chalkboard:

4. Allow students to work individually or in groups. Before they begin, or after a few minutes, **offer some hints** on how to go about the investigation. Write possible strategies on the board (**look for and use patterns, work systematically, record what you do**) and briefly discuss what each might mean in this situation.

5. While the students work, the teacher can circulate, suggesting new lines of enquiry etc. to students who are stuck.

6. (Last few minutes). All students should jot down, for their own later reference, **what they have found out** and **what their current guesses are.**

Lesson 8G2

1. This is a working session with the teacher circulating to help individuals. Draw class together at intervals to discuss common problems and the implementation of strategies. **Students will find an abundance of patterns and the major difficulty will probably be integrating them into one master pattern.** Not all students will achieve this.

2. Suggest possible **extension problems** to students who master the problem quickly. Fruitful avenues to explore are:

    (i) How can you predict the number of different staircases that a staircase number has? For example, 2001 has 7 different staircases - why?

    (ii) What happens if the difference in size of consecutive stairs is not one but two [10 = 4 + 6, 15 = 3 + 5 + 7] or three, or ...?

    (iii) What if the staircases are of a different shape?

3. (Last few minutes) **Again make sure students jot down everything they have found out.**

Lesson 8G3

1. Make it clear how long students have to complete the report and what format it should take.

    Contrary to the cliché, experience is not a good teacher. To learn from experience it is also necessary to reflect upon it. We ask students to write reports in order to encourage (force?) them to reflect upon, and thereby learn from, the experience they have in solving the problem. It is also beneficial for developing vital skills in communication. However, students tend to view writing-up as a chore, so teachers will need to find some motivation for these reports - perhaps a class display for parents' day, a presentation for another class, or something for the school magazine. It is hard to want to communicate if you expect no one will read what you write.

128 *Strategies for Problem Solving*

2. Spend time at the beginning of the session, and throughout, discussing things like
    - what should/could go into the report,
    - the need to make rough copies first and improve expression,
    - the need to be precise, to give examples, etc.,
    - how work can be divided among members of a group.
3. Display the reports.

MATHEMATICAL NOTES

Many interesting patterns can be found amongst the staircase numbers. These can be proved readily using the formulae for arithmetic progressions, but this is not necessary and children will use simpler geometric ideas, or the concept of average or middle numbers. For example, children might see 18 as

$$18 = 3 \times 6 = 6 + 6 + 6 = 5 + 6 + 7.$$

Here are some different approaches often used by students.

(i) By looking in turn at staircases with 2 stairs, then 3 stairs, then 4 stairs etc., many patterns can be found, including

  (a) all odd numbers (except 1) are staircase numbers and have 2 stairs;

  (b) all staircase numbers with 3 stairs are multiples of 3. (In fact, all staircase numbers with a given odd number of stairs are multiples of that number);

  (c) all staircase numbers with 4 stairs are **not** multiples of 4, but instead are even numbers not divisible by 4.

(ii) There are interesting patterns in the sequence of staircase numbers with a given first stair. For example:

  First stair of 1 gives the staircase numbers 1, 3, 6, 10, 15,... .

  First stair of 2 gives the staircase numbers 2, 5, 9, 14, 20,... .

(iii) By looking at the numbers 1, 2, 3, 4, ... in turn and trying to express them as sums of consecutive integers, students observe that **powers of two are not staircase numbers.**

When these results are combined they lead to this discovery - a master pattern.

RECIPE FOR WRITING A NUMBER, x, AS A SUM OF CONSECUTIVE NUMBERS

Step 1: Choose any odd factor of the number except 1. (If 1 is the only factor, then the number is a power of two and so not a staircase number anyway.) Call this odd factor n.

Step 2: If the chosen odd factor, n, is small enough, then it is equal to the number of stairs in the staircase and x/n is the height of the middle stair, (i.e. the average size of a stair).

[Here n is "small enough" if x is at least as large as the smallest staircase with n stairs, i.e. if $x \geq 1 + 2 + ... + n$.]

Step 3: If n is not small enough, then the average height of the stairs is n/2 and number of stairs is 2x/n.

We will illustrate how the "recipe" works, using the example of x = 30.

The odd factors of 30 are 1, 3, 5, 15. We will choose each of these odd factors in turn and apply the steps of the "recipe".

(i) Let n = 1. This is the uninteresting case of just one stair of height 30 - see step 2.

(ii) Let n = 3. Since $30 \geq 1 + 2 + 3$, step 2 applies and we have 3 stairs, of average size 10, giving $30 = 9 + 10 + 11$.

(iii) Let n = 5. Since $30 \geq 1 + 2 + 3 + 4 + 5$, step 2 applies and we get 5 stairs of average height 6, giving $30 = 4 + 5 + 6 + 7 + 8$.

(iv) Let n = 15. Since $30 < 1 + 2 + 3 + ... + 15$, step 3 applies and the average stair height is $7\frac{1}{2}$ and there are 4 stairs. So

$$30 \quad = 7\tfrac{1}{2} + 7\tfrac{1}{2} + 7\tfrac{1}{2} + 7\tfrac{1}{2}$$
$$= 6 \; + 7 \; + 8 \; + 9.$$

Note: The staircase $30 = 6 + 7 + 8 + 9$ can also be thought of as all that is visible of a staircase with 15 stairs with middle stair 2, since

$30 = 9 + 8 + 7 + 6 + 5 + 4 + 3 + 2 + 1 + 0 + (-1) + (-2) + (-3) + (-4) + (-5)$.

$= 9 + 8 + 7 + 6.$

So it really does come under Step 2, as well.

Reference: **"Thinking Mathematically"** (Chapter 4), J. Mason, L. Burton, K. Stacey, Addison-Wesley, London, 1982.

Lesson Block 8H

WHAT IF...?

Theme   The benefits of exploring "What if...?".

Aims   To give students practice in articulating patterns in a situation where patterns are easy to see but hard to say.

To encourage students' curiosity and creativity when exploring their own "What if...?" ideas.

To show how investigating "What if...?" questions illuminates the original situation, by distinguishing those features special to it, from those that are more general.

Mathematical Activity

DIAMONDS

Take a large white square and mark the midpoints of the sides. Join the midpoints and colour the resulting square black. Join the midpoints of the sides of the black square and colour the resulting square white. Repeat this process.

Explore the patterns created.

STEP 0   STEP 1   STEP 2   STEP 3

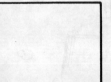   

Many variations of this simple situation can be explored.

Summary of Lesson Block

Lesson 8H1   Looking at the original DIAMONDS situation, finding patterns involved, articulating them clearly and, wherever possible, explaining why.

Lesson 8H2   Exploring some variations of DIAMONDS and seeing how the patterns observed earlier change. Articulating these changes and explaining why they happen.

Materials Required and Teacher Preparation

All students will need rulers, pencils, rubbers and coloured pens. **Templates for geometric shapes** may save time when doing some of the variations. Graph paper could be used to save measuring in Lesson 8H1.

Teachers should choose their own favourite "What if ...?" situation and explore it, noticing how the results compare with the original DIAMONDS situation.

SAMPLE LESSON PLAN

Lesson 8H1

1. Introduce the DIAMONDS situation at the chalkboard and ask the students to make their own copy up to at least step 3. Explain that the process can be continued indefinitely.

2. Ask the students to suggest questions to investigate or to make guesses about the DIAMONDS situation.

    The teacher may need to begin this process with one of the suggestions from the Mathematical Notes (for example "What is the colour of the middle region?" or "How many 'squares' ☐ and how many 'diamonds' ◇ are there?").

    At this stage, many of the responses will probably be too vague to record on the chalkboard. It will be necessary to engage in a dialogue with the student to establish exactly what he or she means. For example, the student who suggests counting diamonds needs to make precise what is meant by a diamond - is it any square or just a square "on its side"?

    Encourage the rest of the class to participate in these dialogues until the ideas are clear, then record them on the chalkboard.

3. During this process make the point that it is frequently not easy to express mathematical ideas clearly. Often it is **easy to see** a pattern, but **hard to say** exactly what it is. Similarly it is frequently **easy to get a feeling of why** something is true, but it is very much **harder to express your reasoning clearly** so someone else can understand. It takes some solid thinking to convert a sense of what is going on into a clear statement. No-one has a formula that makes it easy to express things clearly. The basic technique is to accept that it requires thought and to improve your first attempts by re-drafting, perhaps several times.

4. Remind the students that the DIAMONDS process can be continued on as long as we like. Patterns can be used to help discover things about later steps which haven't been drawn. For example, we could find out what would happen at step 10 or at step 67.

    Select one of the questions arising in 2. above. As a class, find a clear statement of the pattern and a clear explanation of why the pattern works. Then use it to answer the question about step 10 and step 67.

5. Ask the students to select another question of their choice and answer it for steps 10 and 67. Insist students **write down the pattern clearly** and **explain why it is true.** The DIAMONDS problem is useful because the patterns and explanations are fairly easy to see, but hard to use until they have been expressed succinctly.

    [Some students may be fascinated to think about what would happen if the DIAMONDS process was continued indefinitely - what colour is the middle point?]

132 Strategies for Problem Solving

Lesson 8H2

1. This lesson is about **"What if ...?"**. By asking "What if...?", we can not only pose and solve new problems, but often learn a lot about the original problem. Variations on DIAMONDS that are fruitful to investigate include:

    What if three (or more) colours are used in turn, rather than alternating black and white?

    What if the original square is replaced by other shapes, such as other quadrilaterals, triangles or pentagons.

    What if the midpoints of the lines are replaced by other points? (For example by the point one third of the way along the sides, moving in a clockwise direction.)

    Allow students to work on selected "What if ...?" variations of DIAMONDS, insisting again on students producing some **clear written explanations**. Two or more periods may be spent on this. It is best if children work on their own "What if ...?" questions - there is nothing as interesting as the question you pose yourself. Some teachers may like children to present their results as a project or oral report to the class.

MATHEMATICAL NOTES

EXPLORING DIAMONDS: POSSIBLE PATHS

1. **What colour is the middle region?**

    The middle region is white at even numbered steps, black at odd numbered steps. (So, for example, at step 10 the middle region is white and at step 67 it is black.)

2. **How many 'squares' ▢ and how many 'diamonds' ◇ are there?**

    At step 0 we start with 1 square. At every **odd** step we add 1 diamond and at every **even** step we add 1 square. So at step x we get the following:

    If x is even, there are $1 + \frac{1}{2}x$ squares and $\frac{1}{2}x$ diamonds. If x is odd, there are $\frac{1}{2}(x+1)$ squares and diamonds.

    (So at step 10 there are 6 squares and 5 diamonds, while at step 67 there are 34 squares and 34 diamonds.)

    Note that only a few children will express results using algebra - do not force this.

3. **Is more of the area black or white?**

    The teacher may need to explain that the area of each "diamond" is half that of its "square". Children could convince themselves of this by cutting out a square and folding the corners over.

Black area = white area.
So area of diamond = half area of square.

Folded square   Unfolded square

There is never more black. If the process was to be continued indefinitely, two-thirds of the area would become white.

4. **How many white triangles are there at any step?**

   At step x there are 2x white triangles if x is even and 2(x + 1) if x is odd. This is because 4 white triangles are added at every odd step.

   (So, for example, at step 10 there are 20 white triangles, while at step 67 there are 136.)

5. **How many black triangles are there at any step?**

   At step x there are 2x black triangles if x is even and 2(x-1) if x is odd. This is because 4 black triangles are added at every second step, starting at step 2.

6. **How long is the side of the middle square or diamond?**

   At every step the length of the side of the middle square or diamond reduces by a factor of $\sqrt{2}$, although, at best, children will only be able to approximate this. After every two steps the length halves, so by step 10 it is reduced by a factor of $2^5 = 32$.

"WHAT IF ...?" VARIATIONS OF DIAMONDS

1. **Colourful variations on DIAMONDS.**

   What if three (or more) colours are used in turn, rather than just black and white?

   Step 6 for 3 colours

   Notice that with 3 colours, each colour is used at every **third** step. So, for example, the middle region is white at step x if x is 0, 3, 6, 9, etc.

   Similarly, 4 white triangles are added at steps 1, 4, 7, etc.

2. **Joining midpoints on other shapes.**

   What if the original square is replaced by a triangle?

   For triangles, the newly created middle triangle alternates in colour and in orientation. It is always geometrically similar to its parent triangle and its sides are half as long. Three white "corner" triangles are added at each odd step and three black "corner" triangles (upside-down) are added at each even step.

   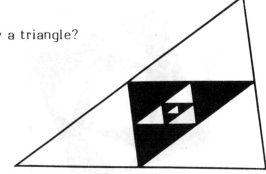

   Step 4 for triangles

What if the original square is replaced by another quadrilateral?

No matter what quadrilateral is used in Step 0, the newly created quadrilateral of Step 1 (and then all subsequent steps) is a parallelogram, with pairs of sides parallel to the diagonals of the original quadrilateral. As with the case of the square, at every second step, the side length of the middle parallelogram is halved.

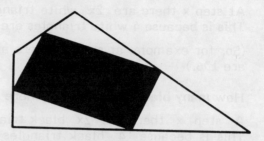

Step 1 for quadrilaterals

What if the original square is replaced by another shape?

Interesting patterns, both geometric and numerical, arise when the original square is replaced by convex or concave polygons.

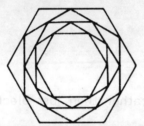 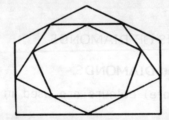 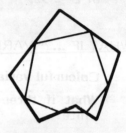

3. **Using points other than the midpoints.**

What if we mark one third of the side of the shape (the first third going clockwise).

For an equilateral triangle the newly created triangles rotate by $30°$ each time, so every fourth middle triangle has the same orientation (and colour) - this requires some knowledge of right angled triangles to prove.

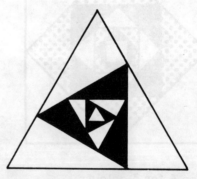

Step 4 for an equilateral triangle

For a square, however, the square is rotated by $26.6°$ at each step, so no such simple pattern emerges.

Step 4 for a square

THE COVER DESIGN

Many attractive designs can be created by varying these ideas. The cover design of this book was created by placing together, in symmetric pairs, six decorated equilateral triangles. Each triangle was decorated in the DIAMONDS fashion, but the mark was made a **fixed distance** along each side.

HELPFUL HINTS FOR PROBLEM SOLVERS

TO START
1. READ THE PROBLEM - REALLY UNDERSTAND IT.

2. MAKE A START - WRITE OR DRAW SOMETHING.
   If you can't see what to do straight away, find
   out what you WANT to do and what you KNOW about it.

AS YOU WORK
3. USE SOMETHING TO HELP YOU.
   Draw a diagram; choose helpful paper; use coloured pens, cubes, counters, scissors and paste, graphs, algebra.....

4. WRITE DOWN WHAT YOU DO.
   Then you won't forget what you did and how you did it.
   Jot down your ideas.

5. NUMBER EACH PAGE.

6. WORK SYSTEMATICALLY.

STUCK?
7. DON'T WORRY IF YOU GET STUCK !
   There are lots of things to do about it. Ask yourself:
   - What do I KNOW about the problem?
   - What do I WANT to do?
   - Can I USE something to help me?
   - Can I make a GUESS?
   - Can I CHECK what I've done?

FINISHED?
8. When you have an answer EXPLAIN WHAT YOU HAVE DONE so someone else can understand.
   CHECK your work.

9. ASK YOURSELF "WHAT IF....?" to get ideas for other problems.